Services and Space

WITHDRAWN

Services and Space:

Key Aspects of Urban and Regional Development

J. Neill Marshall and Peter A. Wood

Longman
Scientific &
Technical

Longman Scientific & Technical
Longman Group Limited
Longman House, Burnt Mill, Harlow
Essex CM20 2JE, England
and Associated Companies throughout the world

Copublished in the United States with
John Wiley & Sons, Inc., 605 Third Avenue, New York, NY 10158

First published 1995

British Library Cataloguing in Publication Data
A catalogue entry for this title is available from the British Library.

ISBN 0 582 25162 1

Library of Congress Cataloging-in-Publication data
A catalog entry for this title is available from the Library of Congress.

ISBN 0 470 23511 X

Set by 7 in Palatino 10/11pt

Produced by Longman Singapore Publishers (Pte) Ltd
Printed in Singapore

for our sanity and future; and for love and support.

to the good and the bad

Contents

List of Figures

List of Tables

List of Boxes

Acknowledgements

With sincere thanks to Betty, Amanda and Charon for their invaluable help in writing this book.

We are grateful to the following for permission to reproduce copyright material:

Basil Blackwell Ltd for extracts from the article 'Restructuring the civil service: reorganisation and relocation 1962–1985' by V Winckler from *International Journal of Urban and Regional Research* 14, 1989, pp. 135–57; Basil Blackwell Inc for extracts from the article 'Flexible Specialization and Regional Industrial Agglomerations: The Case of the US Motion Picture Industry' by M Storper and S Christopherson from the *Annals of the Association of American Geographers* 77, 1987, pp. 104–17 (Editor's note: This abridgement was made by this book's editors and is not the responsibility of S Christopherson and M Storper); The author, Professor W B Beyers, for Table 2.2 (Beyers, 1990); Carfax Publishing Company for extracts from the article 'The Geography of Technological Change in the Information Economy' by M Hepworth from *Regional Studies* 20, 1986, pp. 407–24; The author, Mr T Elfring, for Table 2.3 (Elfring, 1988); European Commission for Fig. 9.1 (Sapir, 1993); The Institute of British Geographers for extracts from the article 'Flexibility in the US service economy and the emerging spatial division of labour' by S Christopherson from *Transactions of the Institute of British Geographers* 14, 1989, pp. 131–43; Longman Australia Pty Ltd for extracts from the article 'Modernity and post-modernity in the retail landscape' by J Goss from *Inventing Places: Studies in Cultural Geography* K Anderson and F Gale (eds), 1992, pp. 159–77; The author, Professor I Miles, for Tables 2.4, 4.2 and 4.3 (Miles, 1993; this table was first published in *Futures*, volume 25, No 6, July/August 1993 and is reproduced with the permission of Butterworth-Heinemann, Oxford, UK); OECD for Table 2.1 (OECD Labour Force Statistics 1992); Pion Ltd for extracts from the articles

'Mrs Thatcher's vision of the "new Britain" and the other sides of the "Cambridge phenomenon" by P Crang and R L Martin from *Environment and Planning D, Society and Space* 9, 1991, pp. 91–116, 'The restructuring thesis and the study of public services' by S P Pinch from *Environment and Planning A* 21, 1989, pp. 905–26, and 'Is the "golden age" of British grocery retailing at a watershed?' by N Wrigley from *Environment and Planning A* 23, 1991, pp. 1537–44; Pion Ltd and the author, Dr S Pinch, for Table 4.1 (Pinch, 1989); Pion Ltd and the author, Dr C W Jefferson, for Table 8.1 (Jefferson and Trainor, 1993); UNCTAD, United Nations, Geneva for Table 5.1 (UNCTAD, 1989).

CHAPTER 1

Introduction

The purpose of the book

This book is designed to support teaching of the geography and economics of services in undergraduate courses on advanced industrial economies, and to show how studies of service activities relate to broader theoretical traditions in the social sciences. It arises from a journey begun in the early 1980s when our research interests moved away from industrial location, and the study of manufacturing industry alone. Services seemed to merit attention, not simply because of the growing numbers of people they employ, but also because of their increasing importance for the success of other economic activities, including manufacturing. Services, we observed, not only create wealth but also mould its distribution and resulting patterns of welfare. Modern service developments are thus of central significance for contemporary urban and regional economies.

Ever since the 1960s, due largely to the pioneering work of Greenfield (1966), Fuchs (1968, 1969), Bell (1973), Stanback and Noyelle (1982), and Gottmann (1983), the impact of service changes on metropolitan economies has attracted increasing attention in the United States. More notable in the UK and Europe, however, was the neglect of services, with only a few exceptions in the 1970s, such as the work of Goddard (1979) and Daniels (1975, 1979, 1982) in geography, and Hill (1977) and Gershuny (1978) in economics. Some services, of course, had attracted a corpus of theory and research, especially retailing, stimulated by Walter Christaller's (1966) seminal Central Place Theory (Berry, 1966). Office functions had been analysed, building upon Swedish research by Thorngren (1970), examining the influence of information exchange on their location. Models of port development had also been proposed (Bird, 1971; Hoyle and Hilling, 1984). But these approaches tended to isolate particular services from wider economic trends, focusing on their spatial manifestations. While offering insights into their changing geography, they were largely silent on the processes driving service

1

growth and change. They also neglected the growing general contribution of services to wider economic change, and all too frequently the employment consequences of service development. Implicit in many commentaries was the view, derived from classical economics, that services are the channels for the distribution and consumption of wealth, which first has to be created in material production. The myth prevailed of a dependent and even parasitic service sector.

There was thus no significant tradition of analysis for addressing the types of regional and urban problem with which we were becoming concerned. For example, research on the impacts of manufacturing restructuring in northern England showed that this increasingly involved changes in the distribution of controlling service workers within manufacturing firms, and in specialist outside service suppliers (Marshall, 1979). Both showed a persistent bias towards growth in the South East of the country. This led to investigation of business service activities themselves, and recognition that they were not simply dependent on local manufacturing, but possessed their own powerful locational dynamics (Marshall, 1982, 1983). Were not service activities, then, taking a leading role in creating regional inequalities?

Another preoccupation at this time was with the decline of the economic base of London's docklands. Changes here were almost entirely a matter of restructuring between service functions, from the old port to new office-based activities. Yet much of the debate about the future of Docklands was concerned with attracting manufacturing to the area, presumably because only this was regarded as offering the basis for a 'proper' economic revival (Wood, 1984).

Reacting to this dissatisfaction, we decided independently, but later jointly, in the Producer Services Working Party (Marshall *et al.*, 1988), to focus on the causes of service growth and spatial differentiation, to create a better understanding of the economic geography of the advanced economies. The Working Party pulled together existing evidence and research on business services and acted as a stimulus for further work. But we were conscious at the time that the exercise included only a limited range of activities, and that other types of service function also deserved more attention.

As is so often the case with academic fashion, our reorientation of interests was not unique, and the last decade has seen a burgeoning of research on service activities. It is no longer necessary to begin any discussion of services research with a complaint about their neglect. Empirical studies have multiplied, and theoretical debate has diversified. The subject of enquiry has also expanded. At first, business and financial services attracted most attention, because of their growth and obvious economic significance. More recently, however, processes of change in all types of service activities have attracted more interest, including consumer and public services in the formal economy, as well as services provided informally, in domestic and local community environments.

Service studies, encouraged by such publications as *The Service*

2

Industries Journal, are becoming increasingly multidisciplinary, involving economics, business studies, sociology, anthropology and contemporary cultural studies, as well as geography. Commentators have also adopted a variety of theoretical and ideological perspectives, from neo-classical economic studies of output and labour productivity change, to Marxist-inspired analyses of the growth and composition of a 'service class' of skilled white collar workers.

The organization of the book

Over the years, our interests in services have been reflected in undergraduate teaching in various ways, initially as part of economic or industrial geography courses, and later in specialist courses on the urban and regional geography of services. A recent survey of economic geography courses in British higher education shows that this process has been mirrored elsewhere (Healey, 1994). As this interest has developed, we have become increasingly dissatisfied with the supply of teaching material on the geography of services. This initially reflected the traditional neglect of services. Within a few years, however, the explosion of interest in services has created the opposite problem. Not only is there now a huge volume of work, but any text becomes rapidly out of date. Service studies including economic, social and cultural perspectives, have become increasingly diverse and difficult to present to students in a way that offers a coherent overview of the field, and its relationships with other research in the social sciences.

The inspiration for this book has come from the demands of teaching. Its priorities are clarification and explanation, rather than adding further to the existing body of empirical research. We combine a textbook on service growth and spatial development, with some elements of a reader, reflecting the diversity of service studies and presenting some key contributions published over the last 20 years. The book adopts various approaches to presenting teaching material.

1. The main *text* summarizes the principal issues and arguments. It presents our own perspective on service change and location, and also provides a commentary accompanying the readings, placing them in context.
2. *'Boxes'* illustrate particular issues or contributions in fuller detail.
3. *Readings* present more lengthy passages from other sources. These present more detailed interpretations or case studies of aspects of service location. For reasons of length and focus, they are not full versions of the original papers which may, of course, be traced through the bibliography.
4. Recommendations for *further study and discussion*, at the end of each chapter are intended to direct and encourage readers:
   (a) to undertake their own enquiries. The *activity* suggestions should stimulate either library or field work, as the basis for short investigative projects, 'consultant's reports' or longer

dissertations. Some may involve careful preparation, with the help of course teachers, to choose sample respondents, to design questionnaires and to analyse replies;

(b) to read a variety of material in a focused manner and *consider* its implications for service development;

(c) to *discuss* significant issues, and write essays which summarize their own conclusions about different aspects of modern service development.

The recommendations for further study require the reader to consider alternative interpretations and perspectives, which are often best explored through group discussion and debate. A short bibliography is provided at the end of each chapter to support such studies.

The book thus seeks to provide teaching material in a flexible manner, while at the same time providing a conceptual framework linking together the various ideas. We hope that the material in the book will offer various types of flexibility, in relation to diverse teaching requirements.

Flexibility in relation to diverse course designs

In a rapidly developing field, the presentation of key contributions provides a series of benchmarks for any course involving the study of services. Teachers elsewhere should thus not have to organize their courses in our way to find the readings or illustrative boxes useful. Our primary criterion of choice has been that the readings should have been of substantial theoretical, methodological or empirical significance for service studies. We hope that this status will be retained, even while service studies continue to develop.

Flexibility in relation to teaching method

The expanding volume of research literature, growing student numbers and limited library resources mean that the need for accessible texts as the basis for student-centred learning grows every year. We hope that the book will provide a resource for diversified modes of teaching, including small group discussions, seminars and debates. The readings and illustrative 'boxes' will give students ready access to key literature. With appropriate guidance for further reading, and effective questions for consideration, however, we aim to encourage students towards their own exploration of more detailed evidence.

A framework of ideas for critical appraisal

Our commentary develops and projects our view of how ideas and evidence about service changes should be organized. Its aim is to present a coherent overview, but not in a prescriptive manner. Set in the context of the wider debate in service studies, it should encourage a critical approach to our views as well as other authors.

A service-informed perspective on structural change

We see the development of service activities as part of broader structural changes in the way the advanced economies work, involving material production, aspects of consumption, circulation and regulation. These structural changes have shaped the geography of the advanced economies, but not in a deterministic way. They create new forms alongside existing spatial configurations, and in turn respond to inherited spatial structures. The readings have been selected to reflect different conceptualizations of the geographical consequences of structural change. The commentaries and chapters outline the different research traditions which characterize service studies and highlight the particular perspective adopted in each reading. This shows the way different schools of thought, with different agendas, have developed, as research on service location has become more established.

Though not presenting new research material, the book aims to be forward looking, and to highlight key questions raised by the growth of service activities, which are likely, for the foreseeable future, to form part of any course on service location. These include the following:

1. Why have some service activities grown, and which are likely to grow or decline in the future? To what extent can service growth substitute for a decline in jobs in other sectors?
2. What sorts of jobs are being created in service activities and who is taking them?
3. What factors influence the growth in demand for different services?
4. Why are some services so spatially concentrated, and is this inevitable?
5. What contribution do services make to urban and regional development?
6. How will new technologies affect the type and location of service work?
7. How should services be analysed? Is it worth studying such diverse activities together? What do they have in common in their contribution to the wider economy?

To answer these questions in *Section I* the book begins by documenting the growth and location of service activities. It immediately becomes clear that services must first be defined and classified in meaningful ways, to make sense of the partial and at times inadequate secondary data on their development. Sense must also be made of their diversity, and the way they relate to other forms of economic activity. Different theorizations of service dynamics have different approaches, both to their classification and analysis. There is a symbiotic relationship between theory, method and empirical research on services. So the book shows the way different approaches to service location define and classify services.

In *Section II*, the book then deals with the critical issue of employment creation in services; and organizational, technical, political and social influences on the geography of service work. Attention is directed to two types of service activity which we neglected in our earlier work, but which have attracted increasing attention since: those supporting consumption, and those provided by public sector rather than commercial agencies. We conclude with an analysis of the role of services in urban and regional wealth creation. Throughout, we maintain that, while service functions deserve growing attention, their analysis should not be envisaged as something distinct from material production, or from wider technical, organizational, social or cultural changes. We do not, therefore, produce a special theory of the role of all services in structural change. While the growing prominence of services in urban and regional economies encourages some to envisage the development of a 'service-led' economy, instead we develop a 'service-informed' analysis of urban and regional development. This recognizes the diverse significance of services in local economies, linked in complex ways to other activities, locally and elsewhere, even up to the global scale, and their sometimes leading role in generating economic growth. They cannot be separated, however, from wider processes of economic and social restructuring which are shaped by the demands of profitable production.

Services offer specialist skills to society. Their value, therefore, depends on how they are used elsewhere, in primary manufacturing, or other service production, or by consumers. In the past, this inter-dependence has sometimes been mistaken for dependence. It should now be widely recognized that service functions are vital to both the effectiveness of production and the quality of life. This is why a volume on the economic geography of services must also address the nature of modern economic organization, and many related political, social and cultural changes.

Conceptualizing service location and change

CHAPTER 2

Service growth and location

Service growth and structural change

Service industries now dominate employment in the advanced 'industrial' economies. In 1990 they accounted for 62 per cent of the jobs in the Organization for Economic Co-operation and Development (OECD) countries; some 61 per cent in the European Community, 59 per cent in Japan and 71 per cent in North America (Table 2.1). Service employment grew dramatically in the OECD between 1970 and 1990. While agriculture declined by 12.4 million jobs, and industrial employment increased by only 1.7 million jobs, service sector employment grew by 89.4 million. North America led this growth. Service employment in the United States increased by 74 per cent between 1970 and 1990, and Canadian service employment grew by 84 per cent. Growth was slightly less dramatic in the European Community, but service employment still increased by 60 per cent. In Japan service employment increased by 54 per cent, and from a small base in 1970 the service sector in Turkey by 1990 had increased by 176 per cent (Table 2.1).

Data cited by Elfring (1988) show marked contrasts in the growth performance of different services in each country. Nevertheless, focusing on the US and the UK highlights important general trends. Table 2.2 shows that in these countries a large share of service employment is in services supplying individual consumer needs. Between 1979 and 1989, these industries made an important contribution to employment growth, concentrated into activities such as hotels, leisure and recreation services. Consumer services, including an important element of public sector provision, generally grew rapidly until the late 1970s (Elfring, 1988), but in the UK between 1979 and 1989 growth levelled off, and employment in the public sector increased by only 14 000 jobs. In both countries between 1979 and 1989, the main contribution to growth came from financial and business services. Though in 1979 they accounted for fewer jobs than consumer services, they grew rapidly in the subsequent decade – by 59 per cent in the UK and 66 per cent in the US.

Table 2.1 Service employment in OECD countries, 1970–1990

	1970 Absolute (000)	Civilian Employment (%)	1990 Absolute (000)	Civilian Employment (%)	1970–90 Growth (%)
US	48 083	61	83 658	71	74
Canada	4 866	61	8 947	71	84
North America	52 949	61	92 605	71	75
Japan	23 890	47	36 700	59	54
Australia	2 964	55	5 418	69	83
New Zealand	523	49	954	65	82
Denmark	1 173	51	1 765	67	51
Finland	906	43	1 489	61	64
Iceland	38	47	75	60	97
Norway	731	49	1 370	69	87
Sweden	2 060	54	3 045	68	48
Austria	1 342	44	1 887	55	41
Belgium	1 892	53	2 570	69	36
France	8 602	47	13 904	64	45
Germany	10 999	42	15 874	57	44
Ireland	450	43	629	56	40
Luxembourg	65	46	125	66	92
Netherlands	2 569	55	4 333	69	69
Switzerland	1 428	45	2 119	60	48
UK	12 686	52	18 305	69	44
Turkey	2 251	18	6 202	32	176
Greece	1 072	34	1 771	48	65
Italy	7 749	40	12 383	58	60
Portugal	1 240	37	2 120	47	71
Spain	4 575	37	6 890	55	51
EC	54 072	45	86 669	61	60
OECD	143 154	49	232 533	62	62

Services: major divisions 6, 7, 8, 9, 0 of the ISIC.
Source: OECD (1992).

A significant characteristic of service employment creation has been the large proportion of jobs taken by women (McDowell, 1992; Christopherson, 1989). In 1965, female workers accounted for more than 60 per cent of service employment in most of the industrial countries, and this had risen to more than 70 per cent by 1985 (see Table 5.1). Much of this employment has been part-time (OECD, 1983). In the UK, for example, between 1971 and 1981, some two-thirds of the increase in private service employment consisted of female part-timers (Robinson, 1985). This enabled female employees to combine work and family responsibilities. As full-time male employment in other sectors declined,

Table 2.2 Change in employment in the United States and Great Britain

Industry group	United States (000)		
	Nov. 1979	Nov. 1989	Change
Extractive and transformative			
Agriculture, forestry and fishery	3 455	3 378	
Mining	985	744	
Construction	4 877	5 499	
Manufacturing	20 974	19 600	
Total	30 291	29 221	–1 070
Distributive			
Transport, communication, utilities	5 253	5 876	
Wholesale trade	5 249	6 307	
Total	10 502	12 183	1 681
Retail trade	15 326	20 021	4 695
Not for profit			
Health Services	5 128	7 827	
Education (private)	1 146	1 772	
Total	6 274	9 599	3 325
Producer services			
Finance, insurance, real estate	5 031	6 849	
Business services	2 947	5 878	
Legal services	473	910	
Membership org. (part)	250 (est.)	234	
Misc. professional	965	1 473	
Total	9 666	15 344	5 678
Mainly consumer sevices	6 021	7 449	1 428
Private households	1 301	1 108	–193
Government			
Federal	2 760	2 966	
State*	3 573	4 288	
Local*	9 583	10 974	
Total	15 916	18 228	2 312
Total employment	95 297	113 063	17 766

* Public education employment: 1979 = 6887; 1989 = 8020.
Source: US Dept of Labor, *Employment and Earnings*
(Data provided by W Beyers, 1990b)

service growth supported a sharp increase in the female share of the workforce in the main industrial economies; from 21–41 per cent in 1960 to 36–48 per cent by 1985 (Table 5.1).

But what are service activities? The conventional definitions employed in these tables group industries with outputs which do not

Table 2.2 Change in employment in the United States and Great Britain (*continued*)

	Great Britain (000)		
	June 1979	**June 1989**	**Change**
Extractive and transformative			
Agriculture, forestry and fishery	356	280	
Mining	335	173	
Construction	1 255	1 035	
Manufacturing	7 015	5 129	
Total	8 961	6 617	−2 344
Distributive			
Transportation services	1 043	899	
Communication, Utilities	762	733	
Wholesale trade	546	935	
Total	2 351	2 567	216
Retail trade	2 642	2 767	125
Consumer services complex	5 120	5 939	819
Producer services			
Finance, insurance and real estate	1 181	1 170	
Business and professional services	476	1 583	
Total	1 657	2 753	1 096
Government	1 580	1 594	14
Total employed	22 311	22 237	−79

Source: UK *Employment Gazette*.

involve the direct production of material goods; sectors which are relatively detached from material production. They include, for example, the education and medical sectors, financial services such as insurance and banking, personal services including shops and other retail outlets, transport and distribution activities and a plethora of services to business clients, such as management consultancies, market research organizations, advertisers and computer services.

These definitions of services have evolved through custom and practice during the past hundred years or so. In the nineteenth century, the attention of economists was focused on farming, mining and manufactured goods production as the heart of the economy. The remaining, 'residual' activities were regarded as concerned mainly with consumption and eventually came to be labelled the service sector (Fisher, 1935, 1939; Clark, 1940). Such a simple distinction has become increasingly unsatisfactory as this sector has grown. If 'services' are merely what is left-over after production has been accounted for, they are hardly likely to form a coherent group of functions with comparable

Table 2.3 Percentage employment by sector in major industrial economies

| | 1870 | | | 1960 | | | 1984 | | |
	Agri-culture	Indus-try	Ser-vices	Agri-culture	Indus-try	Ser-vices	Agri-culture	Indus-try	Ser-vices
France	49	28	23	21	36	43	8	32	60
Germany	50	29	22	14	48	38	5	42	53
Japan	73	na	na	33	30	37	9	34	57
Netherlands	37	29	34	11	41	48	5	28	27
Sweden	54	na	na	15	42	43	5	29	66
UK	23	42	35	5	46	49	3	32	65
US	50	24	26	8	31	61	3	25	72
Average	42	30	27	15	39	46	5	32	63

Source: Elfring (1988).

characteristics. Does it make sense to consider such diverse activities together at all? This is an important issue to which we will return.

Nevertheless, if we accept the conventional definition of services for the moment, recent service growth represents a significant shift in the employment structure of the so-called 'industrial' economies. This has generally corresponded with a move away from 'blue collar' manual production towards 'white collar' office employment. Such change is not peculiar to the last two decades or so. As Table 2.3 shows, services have steadily increased both their relative share and absolute contribution to employment since the last century. But the pace of change has intensified.

One reason for the growing contribution of services to the economy is the decline of manufacturing employment. After reaching a peak proportion of total employment in many industrial countries during the 1960s and 1970s, it has since fallen back in most of them. Consequently, the percentage share of these countries' employment in manufacturing in 1984 was not much different from that in 1870 (Table 2.3). This reversal in manufacturing employment is demonstrated particularly graphically in the UK. From being a 'workshop' economy even up to the mid-1960s (Rowthorn, 1986), manufacturing employment has since declined by more than 3 million jobs. Thus any account of the growing significance of services must acknowledge changes in the manufacturing sector, including rapid growth in its labour productivity.

The growing prominence of modern service activities has accompanied a series of other significant economic adjustments. These include the following:

1. A blurring of the distinctions between goods and service production and a deepening of the links between the two sectors. The production process in manufacturing requires the support of services such as research, design, marketing and selling. Many service industries, on the other hand, such as retailing and banking,

increasingly rely on manufactured products, including computer systems and other microelectronics products, to deliver their output.

2. The growing domination of large private and public sector organizations. These have the capacity to control production and administration ever further across geographical space. There have also been various phases of restructuring in the administration, management and production of these organizations, as they first expanded, and later sought to reduce the costs of their white collar service functions. Private firms especially have developed 'leaner' and more 'flexible' forms of production to buttress themselves against growing competition, with less hierarchical forms of business organization emerging in the 1980s. This has been associated with a resurgence of small firms, especially in service industries, frequently linked to larger organizations.

3. A progressive internationalization of economic activities. This involves the trading of goods and services both for consumers throughout the world, and within and between large companies operating in different countries. The development of international financial service markets has played a key role in establishing this global economy.

4. The widespread adoption and use of new forms of microelectronics technology in the office and in production, combining enhanced computer processing power and communications facilities. New technology thus provides a vehicle for economic restructuring. Established manufacturing and service industries have become more knowledge- and capital-intensive as they have adopted micro-electronics technology. The speed of economic change has been increased, as constant technical innovation becomes a critical component in economic success. Microelectronic technology has also helped to create a global market-place, linking places together and extending the reach of corporate decision making.

5. Major changes in the scale and organization of consumer markets. Businesses are faced with rapidly changing demands for more varied and specialized goods and services. This requires them to be more responsive to consumer trends and tastes, reflected in more sophisticated forms of marketing and selling by companies, again increasingly operating globally.

6. A growing dependence on a skilled, technically proficient and adaptable workforce for industrial competitiveness. Some service activities, providing expertise such as research, design, consultancy and management, are thus at the very heart of successful economies. At the same time, the potential of new technologies threatens to displace much routine service work, including data processing and clerical work.

7. A move in public policy away from Keynesian demand management and its associated supportive cushion of social welfare services. This has been paralleled by a more entrepreneurial style of state activity, promoting efficiency-led policies aimed at successful adjustment to economic change, very much following many of the trends outlined above for the private sector.

Capitalist economies, by their very nature, must continually change. The scale of these changes, and the range of adjustments involved, nevertheless suggest the current period is one of particular economic turbulence (Harris, 1988). Any understanding of the growing prominence of services during the past two decades must recognize these structural adjustments, and that the emergence of service functions with new locational preferences is also a powerful element in modern economic geography.

The next chapter will explore different approaches to understanding service growth and location. First, we explore the various factors influencing service growth and assess the implications for locational analysis of their development.

Why have services grown?

There are two general reasons why employment in many services has increased:

1. Relatively slow growth in service labour productivity compared with other sectors.
2. A more rapid increase in service demand.

Productivity explanations dominate much economic debate. We shall thus examine these arguments, before summarizing key demand trends.

It has been widely assumed in the past that services are relatively labour-intensive compared with manufacturing. The evidence is based on conventional measures of the value of their output compared with labour inputs (Blois, 1985; Leveson, 1980; Kendrick, 1985). On this basis manufacturing is regarded as the technological heart of any economy, and many consumer and public services appear to offer little scope for greater efficiency through automation. As the economy grows, therefore, manufacturing tends to reduce its share of employment by investing in machinery, while services can generally expand only by increasing their workforce. Employment thus shifts towards low productivity services.

In the late 1960s, Fuchs (1968) and Baumol (1967) analysed the aggregate growth of service employment in the US over the long period from 1929 to 1965. Baumol concluded that over half of this growth could be explained by the lagging productivity of services, rather than by demand changes. Less than one-third was due to more 'contracting out' of demand by businesses and consumers, and only 14 per cent was attributed to rising consumer incomes. Lagging productivity seemed thus to be established as the central cause of the historic drift in employment to services (Inman, 1985).

The overarching significance given to labour productivity differentials implies a 'cost disease' in services (Baumol, 1967, 1985). If service workers are to maintain wage levels comparable with those in the economy as a whole, lagging productivity makes services increasingly expensive compared with goods. This high cost of service production is variously taken to limit their growth potential (Fuchs, 1968), to be a 'drag' on the performance of the economy as a whole (Bacon and Eltis, 1978), or to act as a stimulus for new forms of service provision (Gershuny, 1985). Rising real costs will in time reduce the growth of demand in both the private and public sectors, and service employment growth will also be constrained or even reversed by any attempts to increase productivity. Alternatively, of course, many service workers will be condemned to relatively low and declining wages.

The productivity-based interpretation of service employment change has value in warning of the possible limits to service growth. Nevertheless, the inference that service employment growth has mainly been a consequence of their relatively poor economic performance, and that this bodes ill for the future of the economy as a whole, is debatable. The issues particularly focus on (a) the diversity of service activities, and (b) assessment of their economic role simply in terms of their own labour productivity.

Service diversity

The diversity of services, and their potential for productivity change, has become increasingly recognized over the past 20 years. Baumol himself has drawn a sharp contrast between what he calls 'stagnant' and 'progressive' services (Baumol, 1985). The first group includes labour-intensive, person-to-person services, such as those provided by doctors, teachers and even live musicians, whose qualities are inherently difficult to standardize, and where employment levels rise or fall mainly in response to effective demand. The second group is exemplified by telecommunications, in which there is virtually no contact between the customer and the producer, and modern productivity improvements far exceed those even in most manufacturing sectors, with a consequent huge reduction in real costs. These services represent the ultimate in modern automation. They also demonstrate the capacity of such innovations, while displacing many jobs in older forms of commun-ications, to create significant new demands, and new employment, for example in telecommunication-based information and sales services (Barras, 1985; 1986; see Table 4.2).

Baumol also identifies a third, 'asymptotic' category of service which combines progressive and stagnant productivity characteristics. Although these are able to sustain major productivity growth for a period, this is self-extinguishing. He cites computational services as the clearest example, in which computer hardware developments have supported massive productivity gains (he suggests a figure of 25 per cent per year compound over 20 years). These very gains, however, reduce the scope for further productivity improvements, as computation services become increasingly dependent on specialist labour inputs, in software and systems design, oriented to the needs of specialist markets or even individual clients. The 'stagnant' element of the service has thus taken over, today constituting over 90 per cent of costs, and reducing the scope for further productivity improvements. A similar dichotomy is suggested in TV and radio broadcasting, combining progressive improvements in transmission technology with a growing dependence on labour-intensive programme production.

Any blanket statement about the relative productivity of services and its employment implications is thus obviously unrealistic. Employment in all three types of service has grown in the past, in spite of their very different labour productivity experiences. Even stagnant, labour-intensive personal services are also not immune to technical change. Manufactured products are improving the efficiency of medical service delivery, teaching (see Chapter 8) and, of course, access to music. Household tasks have been transformed by domestic appliances of all sorts, and by fast foods. The real costs of purchasing manufactured substitutes for traditional services or for 'self-service' have fallen dramatically over the past 30 years (see Chapter 5). The limits to these processes seem to lie only in consumers' attitudes to the quality of new compared with old forms of service delivery, and their capacity to afford the remaining, high-cost personal services. Equally, of course, the 'stagnant' components of 'asymptotic' services, such as computer pro-gramming or TV production, are also being affected by new methods. More generally, the productivity of many traditional personal contact services, such as business and financial consultancy, has been trans-formed by the word processing and document copying revolutions, and by communications improvements, from the fax machine to digital global data networks.

Thus, ironically, in attempting to elaborate the 'lagging productivity' hypothesis, Baumol's categories also reveal its limitations. Radically different technologies for improving service productivity are evidently now widely available compared with before 1965 (Table 4.3). Even if the division between relatively high productivity manufacturing and low productivity services applied in the past, the contrast between the size and relative scope for productivity improvements in the manufacturing sector compared with many services is probably much less than it was. Elfring (1984, 1988, 1989) indicates that today less than half of the service employment increase is due to slow labour productivity growth. We shall suggest in the next section that demand patterns, both for financial and business services and consumer services have become more dominant.

Perhaps most significantly, in acknowledging service diversity, Baumol also reveals that there is more to service employment change than comparative productivity pressures. In the case of 'asymptotic' computational services, for example, the full economic benefits of hardware-based productivity improvements (which, of course, reflect manufacturing technology) could never have been delivered unless complemented by labour-intensive software skills. The apparently high productivity of the former cannot be isolated from the latter. Only when computer products have become more effectively aligned by service staff to client needs have they begun to acquire their full economic value. The same even applies to 'progressive' services, such as tele-communications. Their apparently unlimited scope for productivity improvement has little economic impact until the users of communications media, whether businesses or consumers, apply the technology to much more labour-intensive purposes, such as enhancing their efficiency or quality of life, or creating new products. The productivity of any economic function measured in isolation seems therefore to have little significance for its wider employment impacts. As we shall argue throughout this book, the conventional separation of manufacturing and service functions misrepresents economic reality, and especially the significance of service employment for wider developments in productivity and competitiveness.

Service productivity and effectiveness

Quantitative measures of labour productivity were developed first for manufacturing, where the market value of a physical product of defined quality can be compared with the costs of labour required to make it. Such measures are more difficult to establish for services where many outputs are intangible and where outputs in the public sector may not even have a market value (Barras, 1983). This results in proxies being used for service outputs (most notably employment) which devalue analysis of output and especially labour productivity change.

More fundamentally, however, in the case of services, as we have seen, this approach misrepresents their economic role. Riddle has forcefully presented this criticism, citing an OECD report, in 1977:

> the traditional concept of productivity isolates the productive process from the social setting . . . it implicitly assumes closed system characteristics. . . . A distinguishing feature of the service sector is the difficulty, and in many cases, impossibility of considering the production process as a closed system. (Riddle, 1986: 70)

A United Nations Council for Trade and Development (UNCTAD) report of 1984 makes a similar point: 'Efficiency . . . depends to an ever-increasing extent upon the interlinkages which are established among the different productive activities and not only on the productive conditions in the activities themselves' (quoted in Riddle, 1986: 68). Thus, service production depends on critical interlinkages with other economic and social functions.

The output of services is affected by constant change in these patterns of interlinkage. Apparent growth in employment in a particular service, for example, may reflect higher prices or a growing volume of demand, justified by product innovation. Innovation is difficult to measure for services because it often depends much more on the quality of labour skills, and how they are organized, than on capital investment. 'Value added' in services reflects much more the terms of exchange between provider and recipient, than the costs of producing them. The fact that, for example, any two delivery services, airlines cabin procedures or hand-washed car services have received the same investment does not guarantee equal success. Service growth may also reflect a transfer into the service sector of tasks formerly undertaken elsewhere in the economy, in other firms or households. If such functions are 'low productivity' activities, these shifts will increase the apparent productivity of the client, while lowering it in the service sector. The result may nevertheless be a more efficient 'division of labour', allocating tasks to organizations that can perform them better. The overall efficiency of the economy will thus be improved (Chapter 5). Similarly, increased household demand for outside services, while boosting the number of 'low productivity' jobs, may also increase consumer satisfaction.

A particular problem in applying conventional productivity measures is thus to define the quality of the service product (Stanback and Noyelle, 1988). This often cannot be done in advance, being determined by the client in response to the skills of the deliverer. A constant dilemma for service organizations is that increased labour productivity may actually make them less effective, in terms of client benefits or perception of quality. Increasing the numbers of customers served by a shop assistant, patients seen by a doctor, pupils taught by a teacher, or clients of a lawyer or consultant does not necessarily improve the service offered. In contrast, for any manufactured good, the production process that enhances labour productivity can also be designed to ensure its quality.

Assessing the economic effectiveness of services, and how this may be reflected in employment trends, are important issues in the analysis of modern economies. The conventional preoccupation with labour productivity, however, has diverted attention from the real issue of how services contribute in diverse ways to the effectiveness of other functions. Low labour productivity may explain the growth of employment in some services, especially in the past, but much modern expansion has been because services are increasingly needed, and have been highly innovative in developing new markets.

Trends in service demand

Demand for services is spread throughout the economy, in manu-facturing and other services, in production and consumption, and in the public as well as the private sector. These are, of course, highly inter-related, with many providers serving various markets. Patterns of demand for services are dealt with extensively in subsequent chapters; however, in summary, they may be categorized in the following way.

Business demand for services

The structural changes described earlier include a growing involvement of service inputs in the production process, through design, research, consultancy, marketing, financial and computer services (Tschetter, 1987). The following factors are important in explaining the growth of these services (Ochel and Wegner, 1987; Coffey and Bailly, 1992; Coffey and Polese, 1992):

1. The emergence of new goods and service 'products', requiring specialist service support.
2. Transformations in the way goods and services are produced, arising from process innovations which increase specialist service demand.
3. Increasingly complex and internationally integrated financial, production and distribution environments which require additional service support.
4. Changes in government regulation and intervention, invariably requiring businesses to monitor and analyse changes.
5. The proliferation of tasks related to the internal management and administration of firms, especially complex multinational businesses.

Firms may acquire such financial and business services, either within their own organization, or by contracting out to specialist service industry suppliers. During the last decade or so, changes in business organization, including the introduction of 'leaner' and more flexible forms of management, have encouraged firms more towards the latter practice (Rajan, 1987a; see also Chapter 6).

Consumer demand

The growth in services supplying individuals, in sectors such as leisure, entertainment and tourism, has also made an important contribution to the expansion of service employment. The mechanisms behind this growth are the subject of debate, and many feel the arguments for demand-led growth in consumer service employment are less convincing than for business-related services, especially since manufacturing goods have displaced many consumer service jobs (Gershuny, 1978).

Nevertheless, cross-sectional evidence suggests that richer households have a larger share and more diverse range of service requirements than poorer households. The fact that some consumer services are luxuries, associated with affluent lifestyles or the right image, is only part of the reason for this. As incomes increase, demand for goods and their associated support services grows, as does that for a range of personal and public sector services (Illeris, 1989a; see also Chapter 7).

Public service provision

In many consumer services, such as education, health care or social services, there is a strong element of state provision, which is paid for through taxation or similar forms of levy; they are thus offered free or at subsidized cost at the point of delivery. The long-term commitment of governments to social welfare provision has supported a growing demand for public services (Heald, 1983; Jessop et al., 1991: see Chapter 8). This reflects:

1. demographic changes, including population growth and the increasing proportion of older people in the population;
2. changes in family composition, including the growth of one parent households;
3. the increasing participation of women in the workforce, who traditionally carried out caring roles in the family;
4. economic changes bringing about a growth in unemployment;
5. rising demands for improved standards of service provision;
6. the regulatory and supportive role of the state in an increasingly complex and interdependent pattern of economic development has also increased demands for public sector employees.

However, public sector growth has become increasingly fragile, because of the reluctance of government and individuals to accept rising tax levels, and public intervention in the economy. Public services have been increasingly rationalized, with neo-liberal moves to 'roll back the frontier of the state', opening up public sector markets to commercial operators.

The reorganization of the financial sector

The distinction between business and consumer demand has limitations, particularly in failing to capture the special dynamics of the financial sector. The globalization of financial markets has been encouraged by the weakening of national regulations designed to control the financial sector, often in the interests of production, and by rapid advances in the communications technologies used to exploit financial information (Thrift, 1990a; Warf, 1989; see Chapter 9). These have fostered both innovation in financial 'products', and the integration of markets, the most notable feature of which has been international flows of speculative 'hot money'.

This produced rapid growth in international finance during the 1980s, which with further deregulation, particularly in the US and the UK, spilled over into demand for domestic financial services, and related accountancy, legal and real estate services (Thrift, 1990a). The increasing competition and resulting 'debt overhang' from excessive lending in the free-for-all of the 1980s, has meant that the 1990s has been a period of consolidation and rationalization in financial services (Gentle and Marshall, 1992).

The limits to service growth

> Service sector employment growth ... is far from being a pre-determined phenomenon resulting from an autonomous evolution towards a post-industrial society. (OECD, 1983: 51)

Though services have grown rapidly during the 1980s, and constitute the main form of employment in OECD countries, there is no uniform shift towards a service-led economy. There are differences in the character of service growth between countries, and differences in the extent

21

to which services are spatially concentrated within countries (see Tables 2.1 and 2.3 and next section).

Nor are the OECD economies post-industrial in the sense that manufacturing is of little economic significance. Services are 'a very necessary and critical ingredient in all economic growth' (Riddle, 1986). But manufacturing still matters, particularly because it provides the bulk of most country's exports.

Nor is further growth in services on the scale of recent years predetermined (Kutscher, 1988). The processes that have driven service growth in the past, including the increasing demand for business, consumer and public services, and innovation and internationalization in the financial sector, are likely to be less strong in the late 1990s than the 1980s. Many industrial countries have been badly affected by recession, and increasing competition and rationalization, frequently assisted by computer technology, are prominent in many service industries. Some argue that the expansion of the 1980s was the result of a series of 'one-off' events, and that high productivity manufacturing will be the only secure basis for employment growth in the future (Thurow, 1989: 199; see Chapters 5 and 9).

Service location

Shifts in the employment structure of the advanced economies towards services are of interest to students of urban and regional development because they have created new, and in some cases startlingly concentrated, locational trends and patterns. Many services display strong tendencies towards agglomeration, so that locational shifts become cumulative. This certainly seems to have been a dominant trend in recent decades.

Contrasting patterns in service location

Given the varied nature of service activities, locational patterns are similarly diverse (Daniels and Thrift, 1987). Many services are less tied than manufacturing to raw materials and more closely linked to markets. The location of retailing, for example, is determined in large part by the income and purchasing behaviour of customers, and variations within urban areas reflect differences in local income (Marquand, 1979, 1983; Daniels, 1985). Although many retailers were traditionally based in central shopping districts within cities, population decentralization and the wider availability of the motor car have encouraged edge of town, suburban and smaller town shopping centres. This reorganization has also been associated with the growing prominence of large companies, and the declining significance of the small corner shop (Wrigley, 1987). Such reorganization threatens the central shopping area in large cities, although redevelopment may improve these areas, and city centres may still offer specialist and niche retailing.

Patterns of tourism, recreation and leisure, often operating at an international scale, influence a range of services including hotels, sporting facilities, museums and the arts, as well as mainstream consumer services such as retailing and eating places. Many of these services have traditionally clustered along coasts, and around retirement centres and areas of outstanding natural beauty (Townsend, 1992). However, urban tourism is being tapped, especially in the US, in local strategies to develop towns and cities as sites of tourist interest (Law, 1992, 1994).

In contrast, physical distribution and freight transport services are linked more strongly to the distribution of industry. Good access to industrial plants, ports and airports for exports and imports is more important than the distribution of population. The nature of business organization is also an important influence. For, example, trends towards reduced stock levels in businesses and requirements for 'just in time' supply, have increased the ties between companies and their suppliers of components and materials (Sayer, 1989).

Public services, which are not driven by considerations of profit in the same way as private sector activities, also have a different geography. Greater importance is attached to equity of access. This means that many public services are more tied to the local population than private services. Government attempts to alleviate economic disadvantage, by expanding public sector work in problem regions, also gives the public sector a dominant role in such areas (De Smidt, 1985; Hudson, 1988). However, public services are not immune from restructuring to reduce costs and improve services, bringing in train new patterns of location.

The concentration of services in metropolitan areas

In spite of such a variety of service location trends, spatial centralization is the dominant pattern across a range of the most rapidly growing financial and business services, including banking and insurance, computer services, consultancy and real estate. The UK presents a very clear example of this trend. Gillespie and Green (1987), show how in 1981 London had 621 000 of these office-based service jobs, 34 per cent of the national total. A further 12 per cent, 218 000 jobs, were based in five other major conurbations. Similarly in 1989, 60 per cent of financial sector jobs were in 20 urban centres, most in major conurbations such as London, Birmingham, Manchester, Edinburgh, Bristol and Leeds (Marshall *et al.*, 1992). London alone had about 15 times the employment of its nearest rivals, Birmingham and Manchester. At a European scale the concentration of business services in London and the South East of the UK also stands out, along with Paris, and Hamburg (Figure 6.1).

For the US, Stanback and Noyelle have charted the expansion of high-quality business, professional and financial services in large cities. Stanback and Noyelle (1982) estimate that 46 per cent of employment in corporate or business services is located in just 20 Standard Metropolitan Areas. They also identify 39 diversified service centres, including centres serving national markets, e.g. New York, Los Angeles, Chicago and San Francisco; regional markets, e.g. Philadelphia and Dallas and sub regional markets e.g. Memphis and Syracuse (Noyelle and Stanback, 1984). These accounted for 34 per cent of the US population, but contained 62 per cent of the headquarters of the 1200 largest national firms in the US, 83 per cent of the headquarters of the top 200 advertising firms, 52 per cent of the medical schools, 41 per cent of the top universities and 40 per cent of the research and development (R and D) laboratories. Such urban concentrations of finance and business services are also highlighted in Fig. 6.2. In the US only 32 economic areas out of 183 had more than the average share.

Office-based services have become central to the development of metropolitan areas, especially as manufacturing has declined or decentralized. Their influence is exerted both directly, through their own employment, and indirectly, through the purchasing power of their employees, stimulating consumer, leisure and recreation services. The factors encouraging this spatial centralization of services in metropolitan areas include the following:

1. Accessibility and proximity to each other.
2. Good physical access to customers and a large range of other local business activities.
3. Readily accessible transport facilities including international airports.
4. A competitive market environment, maintaining the quality of services, and providing the basis for exports to elsewhere.
5. The availability of high-quality telecommunications infrastructure, including both data transfer and voice facilities.
6. High-quality labour, especially that trained and experienced in business and financial service skills.
7. Ample numbers of clerical and administrative workers.
8. Sites accessible and attractive to staff.
9. An ample supply of suitable quality office accommodation, increasingly offering 'smart' facilities to match the requirements of microelectronics office equipment. For some this requirement includes a prestigious, highly visible site.
10. A high quality urban environment, including cultural, social and shopping facilities.

Many of these locational requirements have become available outside metropolitan centres, giving more weight to some of the latent disadvantages associated with central locations. These include the high cost of office sites, especially for expanding businesses or more space-extensive users; the costs of transport congestion, and problems of attracting a high-quality workforce, without supplementing their salary. Outside cities there may also be greater flexibility of design in constructing buildings and an attractive semi-urban or even rural working environment. Key workers increasingly favour such environments, and the generally better transport of smaller urban areas. Decentralization may also actually improve accessibility to clients who have moved out of the city. The widespread modern access to high-quality telecommunications around major cities often enables functions to be decentralized without significant disadvantage.

Service growth outside large cities has also been encouraged by broader population shifts, as well as technological, market and organizational changes (see Chapters 5, 6 and 7). Many smaller towns are also already significant service centres, for example, as tourist resorts and historic towns, as retirement centres, or even financial service centres, such as local savings banks and building societies which have grown to national significance. Services also dominate growth in many new locations such as the retail, office, distribution and recreational parks associated with motorway access.

What, then, is the future of the metropolitan centre? Despite the out-of-town hypermarket, and shopping and 'shed' retailing on the edge of cities, it is widely argued that cities have become primarily centres for consumption rather than manufacturing and commercial production. They now focus on private and business tourism, recreation and specialist forms of retailing, and also absorb much public sector spending (Dawson, 1988; Castells, 1989). The concentration of these interrelated forms of service demand into cities is thus seen as the dominant mechanism behind urban growth (Lash and Urry, 1987; Cheshire and Hay, 1989). Much ambitious urban renewal in the 1980s was based on consumption, media, tourism, leisure and recreation projects. This included dockland recreation schemes, convention centres, support for cultural organizations, such as orchestras and ballet companies, or popular music and garden festivals. Private investment, often with significant public sector encouragement, focused into large-scale office and shopping complexes. Such initiatives aim to encourage local people to spend more money locally, while also attracting tourists and business visitors. In many cases, these often startling downtown developments renew the traditional consumption role of cities (Cheshire and Hay, 1989; Glennie and Thrift, 1992). Lee (1986), for example, shows that from the eighteenth century to the present day London has played an important consumption role, associated with its importance as a financial and government centre.

A painful transition?

As services become more prominent, what kind of economy is evolving? Noyelle, reviewing his extensive research with Stanback, concludes that the growth of services, as manufacturing declines, will cause considerable disruption in many towns and cities (Noyelle, 1984). Noyelle's 'complex, painful and, as of late, rapid' transition to a service-dominated economy is a far cry from the optimistic scenario painted by Daniel Bell (1973), who envisaged an altogether less stressful transition to a post-industrial economy dominated by professional and managerial staff (Chapter 3, page 40).

Noyelle sees service-dominated labour markets characterized by trends towards dualism; an 'increasing bifurcation in the distribution of earnings and in the structure of job opportunities' associated with service employment growth and manufacturing decline (Noyelle, 1984: 13). Stable, full-time, well-paid manufacturing jobs, are being replaced either by highly skilled and waged technical, administrative and managerial work, or by poorly paid, unstable and part-time service jobs, especially for women. Noyelle also notes the new economic geography emerging from the uneven geography of services. In the US, some older northern cities, such as Boston, have developed service functions to replace their declining manufacturing base. Elsewhere, such as in New York, service job creation has not fully compensated for manufacturing decline, and pressures towards service decentralization (Noyelle and Peace, 1991). Some cities, such as Detroit, continue to depend on manufacturing and are, therefore, more painfully affected by the transition. Finally, smaller centres in the US have suffered from 'upward filtering' of key service activities to larger regional centres.

An extensive discussion of the contribution of service centralization and decentralization trends to changes in US cities has followed this analysis (Kirn, 1987; O'hUallachain, 1989; Beyers, 1991; Stanback *et al.*, 1991; Persky, *et al.*, 1991; Savitch, *et al.*, 1993). In smaller national economies, in Europe and elsewhere, concern focuses on the concentration of services close to capital cities, and their under-representation in many peripheral regions. The latter may thus experience de-industrialization without compensatory increases in service jobs. The widening of the North–South divide in the UK, accentuated by service growth in the 1980s, is a good example of this development (Marshall, 1992). The picture elsewhere in Europe is more complex. Although Howells (1988) argues that specialist business services concentrate in core areas of the European economies, Illeris (1989b) and Jaeger and Durrenberger (1991) suggest that, in general, service growth is much more decentralized, reflecting differences in the urban structure and political history of each country. Chapters 5–7 look at these issues in more detail.

Defining, classifying and analysing services

Current statistical systems emphasise older, mature or even declining economic activities and provide only little information on emerging new industries, products and services, new investment and employment. (OECD, 1986: 20)

... the very language of manufacturing v services is becoming obsolete. Economies are simply a tangle of diverse activities, each of which involves different combinations of 'production-type' and 'service-type' work, and yield products [or services] which vary in the particular combinations of material and informational components which they comprise. We need to find ways of classifying activities which are more sensitive to their (changing) composition. (Miles, 1993: 669)

At the outset of this chapter we indicated that the analysis of service location and change usually relies on unsatisfactory classifications of service activities. Our review of the processes underlying service growth indicated many shortcomings in the way services are measured, since this is dominated by employment data classified according to some version of the 'International Standard Industrial Classification' (ISIC) or similar systems. Questions of definition, classification and measurement are important for service studies. They underpin different perspectives on service growth and location, and preoccupy and perplex many reviewers of service research (Walker, 1985; Daniels, 1985: 275-9).

Any classification must of course separate items, in this case economic activities, into categories. There is always a degree of arbitrariness in how this is done, especially when activities are interrelated in many ways. Nevertheless, classificatory systems such as the ISIC reflect views held by those who devised it of what they judged to be significant. During the first half of the twentieth century, primary and manufacturing production were regarded as the central functions of any productive economy. Thus, the ISIC, and complementary national classificatory systems, were devised to represent the principal end-products of these material processing sectors.

Such systems allow standardized data on sectoral employment or output to be collected and compared over time. They, nevertheless, inevitably become out of date. Adjusting such data collection frameworks costs time and money. Change also tends to be resisted because discontinuities create difficulties for comparisons of trends. Official data sources thus tend to give more detail on old-established rather than newer forms of economic activity, a feature which is more acute during periods of rapid economic restructuring.

Because of their reliance on some form of Standard Industrial Classification, official statistics thus present a predominantly sectoral view of the economy, focused on material processing. The ISIC, and its national variations, have become progressively more detailed, and more recent revisions have identified a wider variety of service functions. They still, nevertheless, contain considerably less detail on services than manufacturing. The UK SIC, for example, was introduced in 1948, when for the first time it provided a framework for measuring service activities in government employment statistics. Yet, in the first revision in 1958, when service employment exceeded that in manufacturing, it contained 32 service divisions compared with 100 manufacturing industries. By 1980, service employment was more than double that of manufacturing, but there were still only 102 service divisions compared with 210 for manufacturing. Even *The Economist* protested, 'Anybody analyzing the restructuring of the British economy can still get far more information on the footwear industry (with 50,000 employees) than on business services not elsewhere specified (with 182,000 employees)' (quoted in Daniels and Thrift, 1987: 4).

A further problem with sectoral industrial classifications is that they group firms, and their employment or output, according to the

27

dominant end-product. They therefore ignore the type of work people do. Thus,

> over one third of those employed in manufacturing industries are in service occupations while almost one fifth of those employed in service industries hold manual jobs. Hence, the accountant in the firm which manufactures cosmetics is classified to the manufacturing sector whilst the accountant in the marketing company which provides a service for the same manufacturing firm is classified to the service sector. (O'Farrell and Hitchens, 1990a: 165)

Official statistics may provide complementary data on occupations, which allow the separate identification of various types of white collar or blue collar work, regardless of end-product. These are, however, invariably less regularly produced than sectoral data, and can be compared with industry data only at broad geographical scales. In Britain, for example, a full census is only available every ten years. The cross-classification of occupations and industry is available only for a 10 per cent sample of returns, limiting the possible detail of analysis. Changes in both the occupational and sectoral classifications also makes analysis at more detail than the national level extremely difficult.

Whatever revisions take place in the SIC, giving more detailed service categories, the problems of service measurement using such classifications cannot escape the basic assumption which underlies them; that the end-product of a firm is the best way of measuring its wider economic significance. For services, whose products are by definition intangible, and can often be valued only when combined with other functions, SIC-based data are always likely to be unsatisfactory. As service functions have increased, product-based sectoral classifications are less able to give an adequate reflection of how economies function. As we have already argued, all final material or service products arise from a complex combination of material processing and service skills.

Thus the notion of a *separate* 'service sector' is an arbitrary outcome of classification procedures designed for other purposes: it represents a 'chaotic conception' (Castells, 1989: 130). The only justification for using such a classification is that it draws attention to the scale of activities within what was formerly a residual category in the economic statistics, otherwise omitted by a preoccupation with manufacturing. The growth of services also reflects important changes in the nature of the modern economy. Nevertheless, for the future we need better ways of measuring economic activity which identify the interconnections of functions, including intermediate as well as final outputs. Good input–output tables, which monitor the demand and supply linkages between industries, are a step in the right direction, though the quality of data on services, and the methods used to collect them are not always reliable (Marshall *et al.*, 1988). However, the contribution of particular activities, which may or may not be particular industries, to these systems needs to be identified. To understand the modern economy, and especially the role of services within it, the dominant sectoral and occupational classifications have to be transcended (Gershuny, 1978).

Reclassifying and reconceptualizing services

The constraints imposed by official data thus run very deep, limiting our ability to identify and measure different types of service in ways that reflect their relationship with other economic activities. Not surprisingly, therefore, in 1956 Stigler argued that there was 'no authorative consensus on either the boundary or the classification' of services (Stigler, 1956). Later, following a 'careful review of subsequent studies' Fuchs also found 'no basis for challenging this conclusion' (Fuchs, 1968). Writing in 1988 we argued, 'Despite a growth of interest in services since, questions of classification and definition remain as problematic as ever' (Marshall *et al.*, 1988: 10). The passage of time has only reinforced this conclusion.

Modern debate over these apparently technical issues of classification has revealed an important division of views over the essential nature of service functions. This is reflected in the next two chapters, distinguishing 'conventional' and 'alternative' approaches to service growth and location. Proponents of the former accept that the service 'sector' is a distinctive group of activities whose growth represents a significant change in the nature of the modern economy. They would, however, wish to refine the basis of service classification by moving away from dependence on the SIC. The alternative view emphasizes how the very separation of services from manufacturing in the SIC has obscured their essential interdependence. Thus, from this perspective, the growth of much of the work which is arbitrarily categorized as services does not represent a real break with past industrial trends. It reflects important changes in the nature of work, but these are still driven by the goals of capitalist production, which consistently seeks out new sources of profit from the exploitation of labour through investment in manufacturing technology. Classification problems are thus not merely technical, but obscure fundamental conceptual disagreements in service studies.

The conventional concern to identify the distinctive nature of services, compared with manufactured goods, tends particularly to emphasize their value to users (Table 2.4). Key features of service outputs are that they:

1. are ephemeral, lasting only for the period of any service trans-action;
2. are intangible or immaterial in nature; and
3. cannot be stored, owned or exchanged.

Many services share these characteristics; the quality and value of a retailing transaction, for example, is brief, and depends on the range of stock and the effectiveness with which its ownership is transferred. Even if a shop's purpose is to transfer a material good, the 'service' content, compared with the alternative of buying direct from the factory, or using mail order, is distinctive, and reflects the particular intangible

Table 2.4 Special features typically attributed to services

Service production

Technology, plant and labour	Low levels of capital equipment; heavy investment in buildings. Some services highly professional (especially requiring interpersonal skills); others relatively unskilled, often involving casual or part-time labour. Specialist knowledge may be important, but rarely technological skills.
Organization of labour process	Workforce often engaged in craft-like production with limited management control of details of work.
Features of production	Production is often non-continuous and economies of scale are limited.
Organization of industry	Some services state-run public services; others often small-scale with high preponderance of family firms and self-employed.

Service product

Nature of product	Non-material, often information-intensive. Hard to store or transport. Process and product hard to distinguish.
Features of product	Often customized to consumer requirements.

Service consumption

Delivery of product	Production and consumption co-terminous in time and space; often client or supplier has to move to meet the other party.
Role of consumer	Services are 'consumer-intensive', requiring inputs from consumer into design/production processes.
Organization of consumption	Often hard to separate production from consumption. Self-service in formal and informal economies commonplace.

Service markets

Organization of markets	Some services delivered via public sector bureaucratic provision. Some costs are invisibly bundled with goods (e.g. retail sector).
Regulation and marketing	Professional regulation common in some services. Difficult to demonstrate products in advance.

Source: Miles (1993).

needs of clients. A personal service, such as medical consultation, or hair styling, is usually brief and unique to each recipient, although its effect may of course last for some time. Material transactions may be involved, but the value of the service is important, and relies primarily on the expertise exchanged between individuals. Similarly, business or financial consultancy services exist only once an exchange of expertise has been arranged. Such advice cannot be stored or 'banked', since its nature is embodied in the reaction to it of the client. Computer data banks may, of course, be used, but the specification and interpretation of such information constitute the critical service. Other service transactions may also be associated directly with material products. Advertising or design work, for example, have tangible products, such as posters, brochures or a prototype. The quality of their design nevertheless determines their effectiveness in relation to the needs and reactions of the recipient. Thus, these material goods are a means to a service end.

The often-quoted formal characteristics of services thus derive essentially from the interaction between buyer and seller which is critical to service provision (Table 2.4). These cannot be embodied in material form, although they may sometimes be associated with material exchange. Frequently the buyer and seller have to come together to produce the service, as for dental treatment. Client and supplier interaction is also often critical to the quality of a service, with the well-informed customer helping to 'produce' a better service (e.g. in medical advice, business consultancy, retail purchase). Even when interaction is not so personal, for example where a computer system can be used for remote booking and delivery, the value of the service still depends on the training, skills and experience of the staff who design and implement the system. This reflects a further characteristic of many services; their tendency towards labour-intensity. Yet, even for capital-intensive activities such as transport, physical distribution and some financial services, the quality of expertise they offer to clients is still critical to their competitiveness in supporting goods or capital circulation (Table 2.4).

Based on such distinctions, and the underlying processes of direct interaction which services require, 'conventional' service studies (see Chapter 3) have sought to argue that services are different from goods production. They offer a 'positive' rather than a 'residual' definition of service functions. Hill defines services as constituting 'a change in the condition of an economic unit, which results from the activity of another economic unit' (Hill, 1977: 336), and Riddle refers to services as 'economic activities that provide time, place and form utility which bring about a change in or for the recipient' (Riddle, 1986: 12).

However, in practice such definitions have to arrive at some accommodation with the sectoral classification of services in official statistics. They accept that such groupings have some notional reality as a result of custom and practice. Thus, Miles refers to services as 'those *industries* which effect transformations in the state of material goods, people themselves, or symbolic (information)' (Miles, 1993: 656, emphasis added). Sectoral data are often regrouped to derive new categories representing their diversity and links with other activities. The most obvious distinction, represented imperfectly in the SIC, is that between *public* and *private* provision. Other divisions of private services have been made on the basis of the types of market served. The aim here is to reflect the essential variety of provider–client interaction, enabling different demand relationships to be more clearly identified. The most common distinction made is between '*consumer* services', such

Table 2.5 A composite definition of producer services based on the SIC (Standard Industrial Classification)

Transport and communications	Physical distribution	Finance
Road haulage	Wholesale distribution	Insurance
Sea transport	Dealers in materials	Banking
Miscellaneous transport		Related financial services

Business services

Property owning and managing
Accountancy
Legal services
Research and development
Professional and technical services
Advertising and market research
Corporate head offices
Other business services (including computer and management services)

Source: Marshall *et al*. (1988).

as retailing, entertainment and personal services, and '*producer* services' supplying businesses. As well as *material handling* freight transport activities, or repair and maintenance, these producer services include such white collar '*business* services' as accountancy, market research, training or management consultancy, and a range of financial services (Table 2.5).

Another approach defines an '*information* sector', highlighting the role of many service activities in producing, processing and communicating information. This usually employs occupational data to identify service workers outside the service sector, especially white collar staff responsible for information collection, creation, manipulation and interpretation, in planning, management and administration functions in all sectors. Similar data have been used to reveal shifts in the internalization and externalization of services by manufacturers. Such classifications attempt to identify occupations which play a critical role in what are believed to be the central tasks in modern economies, information manipulation (Porat, 1977; Hepworth, 1989; Table 2.6).

These recategorizations improve the insights available from SIC-based official statistics. Nevertheless, they do not provide consistent definitions and classifications of services. In the categories labelled as services in the SIC, service provision and goods delivery are intertwined. This leads to arguments about whether capital-intensive transport and physical distribution industries, or even utilities and construction, should be regarded as services. The classifications also cannot escape the lack of detail in the sources upon which they draw, and the problems of viewing firms simply as 'producers' of single end-products. The internal structure of accountancy firms, for example, reflect their role as major collectors, generators, analysers and

Table 2.6 An inventory of information occupations

Information Producers
Science and technical, e.g. chemists and economists; market search and
coordination e.g. salesman and buyers; information gatherers, e.g. surveyors
and quality inspectors; consultative service, e.g. accountants and lawyers;
health-related consultative services, e.g. doctors and veterinarians.

Information processors
Administrative and managerial, e.g. production managers and senior
government officials; process control and supervisory, e.g. factory foreman and
office supervisors; clerical and related, e.g. office clerks and bank tellers.

Information distributors
Educators, e.g. school and university teachers; public information
disseminators, e.g. librarians and archivists; Communications workers, e.g.
newspaper editors and TV directors.

Information infrastructure
Information machine workers, e.g. computer operators and printing pressman;
postal telecommunications, e.g. postmen and telegraph operators.

Source: Hepworth (1989).

interpreters of critical business information. The scale and distribution
of these different functions are, however, nowhere reflected in official
statistics. Even more critical is the lack of information on markets. Like
many other services, accountants supply both business and consumer
markets. Their wider economic significance as a service depends on the
share of output, employment and profit dependent on these market
sectors. Accountancy firms themselves are acutely aware of these
market differences, of course, but they are nowhere reflected in any
generally available statistics.

The financial sector relates uncomfortably, as we have seen, to the
producer–consumer distinction and to an information-based classification.
It, of course, serves both consumer and producer markets, and purveys
information. But it is primarily a critical facilitator of the circulation of
all goods and services through the manipulation of capital linking
producer and consumer functions. With the huge growth of financial
investment capital and of specialist and speculative financial markets,
the success of financial services these days is based much more on the
returns on such investments, or on commission for managing them,
than on income from providing 'service' functions to other sectors,
based on information and consultancy roles.

A more radical, 'alternative' approach to service classification arises
from the Marxist forms of analysis which we shall review in Chapter 4.
Production, the industrial base and its outputs in the form of physical
goods and related labour, is distinguished from circulation functions,
including flows of money, commodities and property rights.
Consumption services support both personal and social (public sector)
consumption (Table 2.7). This classification was originally conceived on

33

Table 2.7 A Marxist-based classification of services

Services involved in goods production
'Services' involved in the production of goods which ultimately have a material output. This includes workers indirectly involved in goods production, i.e. non-production workers in manufacturing industry as well as their business service suppliers, e.g. administration, pre-production activities such as research, design, consultancy etc., and post-production activities such as repair and maintenance. The definition of goods production is wide-ranging, i.e. the production of a legal or consultancy report, or scientific research where the output is on paper are goods production. The cinema is the consumption of a good, as are hotels or 'fast food'.

Circulation
The transfer of goods, labour, money and information, e.g. finance, transport, communications, telecommunications, wholesale distribution, retail trade and property services.

Labour services
These are complete labour processes which do not produce a physical output. This would include waitressing and various forms of advice where the output is intangible and unique, e.g. medical, legal or other forms of professional advice. Sales staff involved in retail outlets, theatre and concert performances, domestic service and teaching.

State functions
Central and local government activities.

Source: adapted from Walker (1985).

the basis that services outputs do not contribute directly to surplus value (profit) or the accumulation of capital because they cannot be stored or owned. Services are necessary, but are secondary to industrial production, and are seen very much in a supporting role (Chapter 4).

While 'conventional' approaches endeavour to define services separately from goods production, this 'alternative' perspective takes an inclusive approach to the definition of goods. To Walker (1985) many 'service industries' are involved in goods production. Computer services, for example, are based on goods because software must be provided in some material form to be traded, for example on a disk. The workforce skills and expertise used in the design, production and maintenance of software and computer equipment have productive value only when they gain such a material expression. Advice and consultancy must also be incorporated in a good, such as on a piece of paper or in a report. The interpretation of services in this way means that not only are utilities defined as goods producers but also, for example, restaurants and other food outlets. These are not regarded as consumer services, but as part of the goods-producing sector; the product, a meal, is a good. The service element, in the quality of cooking and its delivery, can be dispensed with in automated outlets, through self-service techniques. They are thus merely labour-intensive forms of production.

Media activities such as television or film production, and the publishing of newspapers, magazines and books, or films, are clearly dependent on various forms of goods, however 'service-intensive' they may be. Transportation facilitates the circulation of economic activities, especially the movement of goods, using a common infrastructure. Passenger transport supports production by enabling workers to get to work, or consumers to get to the place of consumption. The telecommunications infrastructure similarly provides information and communications functions, which are economically justified fundamentally through goods production.

'True' services are thus confined to what Walker calls 'labour services', which are complete labour processes supplied directly to consumers (see Chapter 4, Box 4.1). Their output is intangible, personal or unique to the user (e.g. a theatre performance), and not normally reproducible. This approach thus defines away many service industries. It is employed mainly to challenge the notion that the growing prominence of service employment signifies a transition to a service-dominated economy. Industrial classifications have misrepresented the economic significance of services so that, 'disparate phenomena are haphazardly loaded onto a single overburdened concept, services' (Sayer and Walker, 1992: 56), and 'many classic industries, such as electric power, construction and transportation have been swept into the capacious bin of the service sector, falsely enlarging its size in national accounts' (Walker, 1985: 65).

Adopting a much narrower definition of services emphasizes the significance of industrial production for service development. For example, with business service production allocated to the goods sector, their growth emerges as a widening and deepening of the social and technical divisions of labour surrounding goods production, and as a part of the wider and longer-term processes of industrial evolution (Sayer and Walker, 1992).

By emphasizing the dependence of many services on goods production, such an approach may over-emphasize the importance of material technology in determining their value. 'Conventional' studies, on the other hand, more readily acknowledge the importance of less tangible workforce skills, expertise and advice, as the key to buyer–seller transactions in many service industries (see page 44), such as computer and professional services consultancy, accountancy or market research, even where material objects are also involved in the production of their output.

The definition and classification of services is thus clearly not a neutral, technical matter. It embodies views, some of which are 'historically-based' in the origins of the SIC, obscuring the modern economic role of services. In examining the construction of a more appropriate framework for the definition and classification of services, a conflict of interpretation has emerged. Those who believe services to be driving new forms of economic production tend to accept to some degree the validity of the categories provided in the SIC, and base their work on their development. Those who argue that service growth represents no more than an elaboration of earlier industrially based developments under capitalism adopt a much narrower definition of what constitutes a service than is provided by the SIC, arguing that many 'service industries' in the SIC are implicated in goods production and circulation. It is these differences of view, and uncertainty concerning the reasons for the growing prominence of service employment in the economy that will be the subject of Chapters 3 and 4.

Further discussion and study

Guided reading

Gershuny J 1978 *After industrial society: the emerging self-service economy.* Macmillan, London: Chapter 4

Hepworth M 1989 *Geography of the information economy.* Belhaven Press, London: Chapter 2

Hill T P 1977 On goods and services. *Review of income and wealth* **23**: 315–38

Marshall J N 1991 Producer services. In Healey M (ed) *Economic activity and land use.* Longman, Harlow, Essex: 362–92

Marshall J N *et al.* 1988 *Services and uneven development.* Oxford University Press, Oxford: Chapter 2

Riddle D 1986 *Service-led growth: the role of the service sector in world development.* Praeger, New York: Chapters 1–4

Stanback T M, Noyelle T J 1988 Productivity in services: A valid measure of economic performance? In Guile B, Quinn J B (eds) *Technology in services: policies for growth, trade and employment.* National Academic Press, New York: 187–211

Activity

1. Choose some familiar domestic needs, such as particular foods, transportation, entertainment, clothing or shelter. Trace the 'chain' of production and delivery that brings them from their origins as raw materials to the consumer.

   Analyse how goods and service expertise are combined to do this.

2. Look at the classification of 'industries' used in a national census, and find out exactly what is included in the detailed ('four-digit') categories of service activity.

   Attempt to group the categories according to whether they primarily;

   (a) are engaged in information or materials processing;
   (b) serve other businesses or personal consumers;
   (c) have a short-term or lasting effect on customers;
   (d) are offered by the private or the public sector;
   (e) support goods production, circulation, or personal consumption.

Is it easy to identify any of these, and which of these exercises is most valuable?

3. For any small town, village or suburb you know, find out where are the nearest: supermarket; post office; specialist hospital (e.g. carrying out cancer treatment); business accountant; bank branch; source of large commercial loans; symphony orchestra; live commercial theatre; university; headquarters of firm with a local branch.

Are these all in the same, or different places? What factors do you think account for differences in their locations?

Consider

1. When customers

   (a) purchase an automobile,
   (b) book an airline seat,
   (c) use a credit card,
   (d) take away a fast-food hamburger,
   (e) subscribe to cable channels on TV,

   what features of the 'product' or 'service' are most important in the decision to purchase?

   Are the material or the service qualities of each more critical?

   When can such service activities be regarded as either a personal or business service? Why does this distinction matter?

2. What impacts have automation had on the delivery in recent years of:

   (a) hairdressing,
   (b) medical care,
   (c) teaching,
   (d) or any other labour-intensive personal service?

   What scope is there for further automation in the future?

Discuss

1. Which services are unevenly distributed, and at what spatial scales? Is the geographical distribution of many services becoming more uneven?
2. If cities are now primarily centres of consumption, where is the money earned to support them?
3. Can service jobs replace those lost in mining and manufacturing?
4. Assess the arguments for both labour productivity and demand-based explanations for service growth.

CHAPTER 3

'Conventional' approaches to service growth and location

> Shifts on the consumption side, coupled with changes in production, information gathering and financing, seem to underlie a remarkable proportionate surge in service employment...any account of the transformation of the advanced capitalist economies since 1970 has to look carefully at this shift (Harvey, 1989a: 156–7)

> There are serious doubts whether a general and consistent theory of structural change exists which is able to explain the complex shifts towards services and the many interrelations in the process of economic development. (Ochel and Wegner, 1987: 24)

The growing prominence of service activities poses a challenge for studies of urban and regional development which, until recently, have generally given priority to manufacturing as the primary engine of economic growth. These studies must not only accommodate the growing dependence of manufacturing on service activities, but also the possibility in some circumstances that some types of service may lead local economic change, thus producing distinctive patterns of regional development.

Studies of service growth and location in the advanced economies at first mainly documented the growing prominence of service employment in national, regional and local economies (Daniels, 1975, 1982, 1985). They did not generally challenge the primacy of manufacturing change for economic development. As manufacturing employment decline persisted and spread, however, anxious attention began to be paid to the contribution to growth and incomes of service functions. Was it positive, negative or merely passive? A debate ensued amongst economists, geographers and sociologists, throwing various perspectives on current change in allegedly 'industrial' economies, including the role of services, and their geographical consequences (Bacon and Eltis, 1978; Blackaby, 1979; Gershuny and Miles, 1983; Wood, 1984; Pahl, 1984; Inman, 1985; Urry, 1987; Marshall *et al.*, 1988; Daniels, 1991a,b; Daniels *et al.*, 1993).

These responses to service growth to some extent reflected the periods when they first arose. Up to the early 1970s mainstream analyses tended to be optimistic about the transition to a service-dominated society. Highly developed service activity was viewed as the hallmark of a healthy industrial economy (Bell, 1973). The recessions of the 1970s and early 1980s, marking a break in the post-1945 phase of economic expansion, undermined this optimism. More pessimistic and critical forms of analysis were advanced, emphasizing the erosion of the manufacturing base, and the dependent nature and low productivity of much of the service employment that replaced it. Marxist-derived commentaries, in particular, offered a critical alternative view of service growth in 'late industrial' society, as a symptom of division and potential crisis (Mandel, 1975). In the 1980s, more detailed document-ation of economic restructuring and a more developed debate have supported a greater cross-fertilization of ideas.

In this chapter and the next we shall review various approaches to the analysis of service location in relation to wider debates in the social sciences. Any description of a diverse field such as service studies clearly involves some simplifications and personal judgement. Readers should explore the literature for themselves, and make their own evaluations. Although this chapter and the next emphasize differences of perspective, the concerns and interests of many commentators overlap. In our view, a 'service-informed' perspective on the conse-quences of modern economic change, building on the strengths of existing research, resolves some of the apparent conflicts of inter-pretation. This does not separate service trends from other aspects of modern economic change; they are too deeply interrelated for this to be sensible. Nor do we give priority to either manufacturing or service functions in explaining change. Services are central to modern economic restructuring and growth, but they play many different roles, in different economic and geographical circumstances. In some cases they follow, and in others lead, wider patterns of restructuring. Nevertheless, their distinctive characteristic lies in the expertise they supply to assist the manipulation of materials, information, capital and labour. The complexity and diversity of much interaction in modern services also encourages their spatial concentration. We shall explain this perspective more fully at the end of Chapter 4. First, we present the most distinctive established perspectives; what we term the 'conventional' approaches to modern service studies.

'Conventional' approaches

'Conventional' approaches to the analysis of services in advanced industrial economies draw on a long tradition of service research (Delauney and Gadrey, 1992). Early economic studies, for example, go back at least to the period before the Second World War when Fisher (1935) and Clark (1940) emphasized the distinction between primary (extractive), secondary (manufacturing) and tertiary (service) sectors, and discussed the growth of the last in the industrial economies (Fuchs,

1968; Delauney and Gadrey, 1992). As the service sector became more prominent, and its diversity more obvious, this simple sectoral approach failed to account adequately for the different roles played by service activities. Attempts were made, especially by Gottmann, to distinguish 'quaternary', information-based, services, and even 'quinary' innovation-oriented activities, from more routine services (Gottmann, 1983). Other distinctions were also suggested, between producer, rather than consumer, services (Browning and Singelmann, 1975; Singelmann, 1978), or information-handling, as opposed to material-handling, activities (Porat, 1977).

Such approaches explain long-term shifts towards a service-dominated society as the consequence of a general increase in wealth, which brings changes in demand. Service growth is seen primarily as a response to shifts in consumer and, more latterly, business demand. As average incomes per person rise, more is spent on more expensive goods, public infrastructure and consumer services. Manufacturing becomes more technically and organizationally complex, emphasizing research and development, and employing a more educated and skilled workforce, and more elaborate sales and marketing strategies (Greenfield, 1966; Gershuny and Miles, 1983). Competition requires costs to be kept down, so that more capital-intensive methods are introduced, gaining the benefits of economies of large-scale production. The proportion of workers directly engaged in goods production then declines, but the share of supporting service work increases, for example, in research, planning and sales functions. The outcome is a technologically advanced, information-based, high wage, high expertise-based society.

Perhaps the most influential presentation of this perspective came in 1973 from Daniel Bell, in his vision of the 'post-industrial' society. He examined the social implications of the emergence of a knowledge-based, educated, white collar 'service' labour force, associated with new patterns of consumer demand, and high discretionary spending. The post-industrial economy was thus,

> a game between persons. What counts is not the raw muscle power, or energy, but information. The central person is the professional, for he [sic] is equipped, by his education and training, to provide the kinds of skill which are increasingly demanded in the post-industrial society. If an industrial society is defined by the quantity of goods as marking the standard of living, the post-industrial society is defined by the quality of life as measured by the services and amenities – health, education, recreation, and the arts – which are now deemed desirable and possible for everyone. (Bell, 1973: 127)

The immediate critics of Bell's optimistic vision pointed out that it was much too idealistic. How would the wealth required to support the high standard of living be created? Although a professional élite might benefit, if a service-based society was to be affordable, it must be supported by a range of service jobs which were more poorly paid and insecure than much traditional manufacturing. The post-industrial society could thus also be a deeply divided society (Kumar, 1978).

Other commentators interpreted increasing service employment in a very different light, raising doubts over the competitiveness of service-based, compared with manufacturing-based, economies (Baumol, 1967, 1985). These pessimists drew on the evidence that an increasing share of service work arises largely from its low labour productivity compared with material production (Fuchs, 1968; see Chapter 2). For comparable levels of output, services employ more workers, and invest less in cost-saving technology. They can do this because service markets are much less open to international competition than those for manufactured goods. There are thus weaker incentives to become efficient. This is especially true for 'non-market' services provided through public or voluntary agencies rather than by private companies (Bacon and Eltis, 1978). This perspective reflected US and European anxieties over the decline of the technological core of the productive economy, predominantly in the manufacturing sector (Blackaby, 1979; Cohen and Zysman, 1987). To survive, this must be internationally competitive, especially with manufacturing in Japan and the newly industrializing countries. The growth of inefficient, low-productivity, tertiary employment could only be supported by income from international trade, yet service growth undermined the ability of manufacturing in the older industrial economies to compete with the new by diverting resources into less competitive functions.

An important problem in this debate is that the different roles played by different services are insufficiently clearly distinguished. The efficiency of modern, capital-intensive manufacturing often depends on the support of some types of service function, which are inherently more dependent on labour skills (Ochel and Wegner, 1987). These offer advanced planning, technical, organizational and sales expertise which cannot be fully automated. Even many competitive consumer services require personal attention to a variety of customers, at different times and places, again limiting the degree of automation. From this perspective, the conventional definition of service productivity, neglecting their contribution to the efficiency of other industries, is highly unsatisfactory (Chapter 2, pages 15–19).

The pressures of international competition also affect an increasing proportion of key services, including financial, business and tourist services, which make a significant and growing contribution to international trade (Riddle, 1986). These developments have also been accompanied by significant technological innovations, especially in communications, information processing and transportation. Such 'high-productivity' service innovation nevertheless also often depends on other specialist services, such as management and technological consultancy, new forms of sales and advertising, specialist insurance or translation services, financial intermediaries, or special quality business travel services. More labour-intensive innovation in internationally competitive service 'products' is thus often actually the basis for competing with standardized, automated services (Barras, 1983).

Perhaps the most comprehensive criticism of Bell's approach, and the pessimistic reaction to it, came from Gershuny. He challenged the whole 'sector' basis of much of this thinking, and especially the association of service employment growth with consumption and information processing, rather than with material production. He showed that growing consumer spending had not resulted directly in more service work, but rather the purchase of more manufactured goods. According to Gershuny, spending on personal consumer services, even including public education and welfare, had not increased during the twentieth century. As Greenfield had argued earlier, much service growth reflected changes in the methods of producing and marketing goods (Greenfield, 1966). The myth of the 'post-industrial' society disguises the fact that we increasingly rely on manufactured goods to provide most of our service needs.

Gershuny thus assigned much formal service employment growth to what he termed the 'indirect' producer and consumer services. Producer services, include many of the much-vaunted 'information-processing' functions, supporting production, rather than consumption. The 'indirect consumer services' increasingly support the consumer use of manufactured goods, by providing infrastructure, finance, advisory, installation, and repair and maintenance activities. These include major new sectors as diverse as television programme production, or DIY-based household maintenance.

Gershuny's analysis of the interrelationships between sectors extended even beyond formal employment activities. He was particularly interested in what he characterized as the 'self-service economy'. Gershuny noted that the increasing reliance of consumers on manufactured commodities has been complemented by changes in household, or 'informal' unpaid labour. Instead of personal servants, consumers have washing machines or vacuum cleaners to help them carry out chores themselves. Instead of chauffeurs or bus drivers, they drive their own automobiles. Instead of going out to the theatre or cinema, they watch the television, or play a video. Often, instead of employing workers to repair or decorate their houses, they do the work for themselves, with the support of the local DIY store. Thus many traditional service occupations have been curtailed in recent decades because of the increasing availability of manufactured substitutes, supported by 'self-service' labour. Manufacturing-driven changes in the 'productivity' of household practice have thus permeated consumption and social behaviour. While 'indirect' services have grown, linked to the production and use of manufactured commodities, many other direct personal services have declined.

There are important exceptions to the trends identified by Gershuny. Some personal services have grown in recent years, reflecting patterns of increased discretionary spending. Tourism, recreation and especially public services form the principal basis of these. Gershuny perhaps also originally failed to anticipate the growing participation of women in the workforce. Their absence from the home for long periods limits the possibilities for 'self-servicing' in the household, and actually pushes some functions such as child care and care for the elderly, into the formal economy, where they add to service employment (Illeris, 1989a).

For the better-off during the 1980s, high female participation in the workforce led to a revival of domestic labour, especially in areas with a supply of unemployed, or low-wage female workers. Polarization of the female workforce has thus occurred, between high and low-skilled work, and changes in consumer services reflect subtle changes in the workplace and the household, arising from the changing role of women in society.

Service growth and location

Many commentators have examined the geographical impacts of the spectacular modern growth of what Gershuny termed 'indirect producer services', supporting increasingly complex processes of production. These specialist studies generally lack the breadth of social analysis offered by either Bell or Gershuny. Their most common characteristic, in the tradition of Bell's work, is to regard the growing modern dominance of such services as a significant break with past economic and location trends. For many, the growth of the 'information economy' has been most critical. Others have given priority to the leading role of technological innovation in supporting information exchange. Another thread of research has focused on the manifestations of these developments in the built environment, especially in the growth of the office sector, dominating the cities of the western world. Others again have emphasized the organizational basis for producer service growth in relation to wider trends in business organization, while some studies have focused on the growth and behaviour of service corporations themselves.

An information economy perspective

Closest to Bell's post-industrial vision are those who emphasize the growing prominence, in job and wealth creation, of a distinctive group of activities concerned with the production, exchange and transformation of information. The number of information-processing workers has certainly increased, related to the complexity of modern business and public sector organization and technology. The control of information sources exercised by some such workers is also regarded as a prime source of new economic activity and innovation. The growing significance of information processing is generally seen as a break with the past; 'another phase of economic history' (Porat, 1977: 204; see also Stonier, 1983), or a 'new phase of economic development' (Hepworth, 1989: 7). Hepworth also points out, however, that the importance of information processing is not new. It has always been at the heart of industrial production, distribution and consumption. Developments in national telephone systems, for example, were associated with new methods of scientific management in the early part of this century, and modern communications developments are now allowing the decentralization of these management structures (Beniger, 1986). The modern dominance of information-processing activities arises from a powerful convergence during the 1970s and 1980s of the information-processing capacities of computers and the expansion of digital telecommunications. The resulting breakthrough in the power of information processing and exchange is regarded by many as a crucial shift or a new 'technological paradigm' leading to a wave of economic growth based on a series of product and process innovations (Gershuny and Miles, 1983; Freeman, 1983, 1987; Hall, 1987, 1988).

The information economy is based in the physical architecture of computer networks and systems, facilitating wider economic, social and geographical change. As Hepworth graphically puts it, through the architecture of the computer network it is possible to locate the information economy literally 'on the map' (1989: 63). Thus, computer networks increasingly enable users to assemble and manipulate information from a wide range of sources, and to distribute their activities across a variety of locations. They also facilitate the growth of geographically tradable information, which in turn makes possible the integration of diverse local economies, bringing about spatial shifts in other economic functions.

Office location and metropolitan development

Growing information exchange has been associated with a huge development of office-based activities. These have been increasingly seen as performing a central role in local economic development over the past 30 years. The view of the UK Northern Regional Strategy Team is typical:

> Offices make a vital contribution to the generation of new economic activity, to the search for and development of new processes, markets and management systems. It is generally in offices too that decisions are taken as to how, when and where ideas for new goods and services should translate into action. (Northern Region Strategy Team, 1976: 12)

Office development is also central to the dynamics of metropolitan change. The spatial clustering of offices in the heart of the largest cities has for long been explained in terms of agglomeration economies, based on the range and immediacy of information exchange, and low communications costs (Foley, 1956). In the 1960s and early 1970s Swedish research enabled models of hierarchical communications which support office location to be developed (Thorngren, 1970), and these ideas were also applied elsewhere (Goddard, 1975; Daniels, 1975; Alexander, 1979; Gad, 1979). Office complexes were found to be underpinned by a dense network of local linkages, frequently involving face-to-face contacts, especially for the more strategically important exchanges of information. Information and accessibility were thus seen to be the key to the concentration of offices in metropolitan centres (Box 3.1). Pred argued that

> Large metropolitan complexes offer ... three specialised information advantages that are seldom available to the same degree in less populous

Box 3.1: Gottmann's transactional city

For more than 20 years, Jean Gottmann was the most influential analyst of office location. He noted the growing prominence of information-processing activities in the economy, and linked this to the growth of a quaternary sector of office activities and to new forms of office-based metropolitan development.

His landmark book *Megalopolis* (1961), charts the emergence of an integrated office complex in urban areas in the northeastern seaboard of the United States. Subsequently, especially in the journal *Ekistics* (Gottmann, 1976, 1977, 1979), he drew on European, North American and Japanese evidence, to show that such office centres were connected by information and communication flows, and formed what he termed a 'transactional city' (Gottmann, 1983). In more recent years he also charted the global reach of such city networks (Gottmann, 1989).

Gottmann argued that, in the second half of the twentieth century, the dynamics of urban centrality no longer reflect the attraction of large-scale factory production to the markets, labour force and transport facilities provided by big cities. Blue collar employment in cities was in decline, and office activities had replaced factory employment as both the foundation of urban centrality and as the dynamo of urban growth, leading to a transformation and rejuvenation of large cities (Gottmann, 1974, 1978).

For Gottmann, the information and communications needs of office activities underlie their spatial centralization in large cities. The manipulation, processing and communication of information are the *raison d'être* of office activities. As office activities grow, increasing occupational specialization produces ever more communication, exchange and teamwork between office specialists (Gottmann, 1978). So, despite developments in telecommunications technology which offer opportunities for communications over space, office workers still require a central location offering face-to-face contact with other office employees, and ease of access to other office centres (Gottmann, 1983). Thus, the expansion of city centre tower blocks, with their vast number of office workers, is the architectural manifestation of the needs of office workers for communication and interaction, and the office block has replaced the factory as the symbol of modern urban development, because the economy of large cities is based on transactional networks of office activities.

metropolitan areas and cities. These are ease of inter-organisational face-to-face contacts, business service availability, and high intermetropolitan accessibility. (Pred, 1977: 177)

The elevation of office activities to such strategic significance for both economic growth and urban development could not disguise the fact that many information contacts are quite bureaucratic and routine, and often easily displaced by automated or remote exchanges, for example by the telephone line. During the 1970s, evidence grew of a trend towards office decentralization from city centres (Daniels, 1979; Burtenshaw *et al.*, 1981). This was not just communications-related; office expansion in city centres was also constrained by lack of space, traffic congestion and higher taxes (Alexander, 1979; Bateman, 1985). Suburban or smaller town locations offered lower costs and a captive labour market. Only the more routine activities, however, were likely to be decentralized (Goddard and Morris, 1976), and communications considerations seemed to constrain decentralization of more strategic functions. Relocation was limited beyond a certain distance because savings in rents or labour costs were outweighed by the increased costs of communications and travel for those still required to attend meetings in the centre (Goddard and Pye, 1977).

Metropolitan office location studies possibly reached their most sophisticated expression in the work of Gottmann (1983), describing the 'transactional city', specializing in and connecting office and information activities (Box 3.1). A more recent variation of this theme has been Moss's (1987a) analysis, arguing that the largest 'world cities' function as critical hubs in an international network of information flows.

Information technology and office location

Microelectronics has transformed the ways in which information is processed and distributed since the 1970s, altering the geography of the information economy. The direction of these changes is nevertheless subject to conflicting pressures and interpretations. Initially the potential was emphasized for multi-party telecommunications and the trans-mission of pictures to substitute for face-to-face contact, and undermine the need for office agglomerations. As early as 1970 Abler observed,

> Advances in information transmission may soon permit us to disperse information-gathering and decision-making activities away from metropolitan centres, and electronic communication media will make all kinds of inform-ation equally abundant everywhere in the nation, if not the world. When that occurs, the downtown areas of our metropolitan centres are sure to lose some of their locational advantages for management and government activities. (Abler, 1970:15)

More decentralized forms of business operations would also be possible if work was no longer supervised through personal contact. Tele-communications technology might be employed to keep workers in peripheral regions in touch with the main information hubs (James *et al.*, 1979; Ernste and Jaeger, 1989). In general, the practical impact of 'teleworking' has so far been limited (Holti and Stern, 1986). More significant have been the dispersal of large-scale routine telephone-based services, such as hotel and travel booking systems, and television and mail-order sales centres (Richardson, 1993; see Box 5.5).

Other technological developments, especially the merging of telecommunications and computer-processing capacity, seem to favour the further concentration of key functions into core metropolitan areas. Telecommunication systems now allow the provision not simply of the basic network, but also more advanced information-processing services. These advanced data processing and communications services are frequently introduced first in major metropolitan centres (Gillespie *et al.* 1984). Moss also suggests,

> Telecommunications regulation is gradually leading to the elimination of cross-subsidies and to a new telecommunications infrastructure, characterised by a multiplicity of telecommunications networks, where the choice and type

of service will vary substantially depending upon the size and scale of the market. Most important, the spread of fibre optic systems, 'smart buildings' and national paging systems is already revealing that the emerging telecommunications infrastructure will favour those metropolitan areas with information intensive industries and lead to disparities between urban and rural telecommunications systems. (Moss, 1987b: 12; quoted in Hepworth, 1989)

Attention has therefore been directed to the changing regulatory regime within which telecommunications operate, and its impact on the infrastructure of the information economy. Communications systems are generally becoming more lightly regulated, and dominated by large corporate interests (Box 8.2). Thus the availability of more sophisticated services in areas with the necessary scale and variety of demand increasingly favours core over peripheral areas (Gillespie and Williams, 1988).

As the reading by Hepworth in Chapter 6 shows, service firms can also locate and relocate facilities more flexibly. Computer networks allow them to extend markets, for example, removing the need for service production to locate close to consumers. Information services such as banking, insurance, legal or data processing may be made widely available from fewer locations. Production, distribution and management units may be technically and spatially separated, each located in their most favourable area (Castells, 1985).

'Information economy' approaches tend to give priority to one part of the economy, viewing broader economic trends in terms of its dynamics. The advanced economies are thus often seen as dominated by the changing volume of information exchange. Only limited attention is given to the quality of the information, the specific uses to which it is put or the effectiveness of its impacts. Even then, a somewhat stereotyped, hierarchical and static view is taken of the economic role of information. Analysis also tends to be preoccupied with the microelectronics technology, 'reading off' the potential consequences of change from its characteristics. Such interpretations are nevertheless broadening their scope. Castells, for example, suggests a move away from the sectoral view of the economy common in the 'conventional' literature. In his view the economic structure is 'made up of processes, in which service activities connect agriculture and manufacturing with the consumption of goods and services, and with the management of organisations and institutions in society' (Castells, 1989: 129).

Castells' analysis has forged a link between studies of the information economy and the 'alternative' tradition, discussed in the next chapter, arguing that the combination of high skills in the information economy, and the capacity of the new technology to make work more routine, is creating the conditions for a polarized society (Castells, 1987). Others concerned with the wider development of capitalism, confirm that information technology has a central role in the rejuvenation of corporate profitability at a time of a crisis in the industrial economies (Robins and Gillespie, 1988). On the other hand, the view that corporations influence the way information technologies are used

(Castells, 1989) reflects a stronger emphasis upon the business organization in service studies.

The regional implications of producer service growth

As we have seen, the 1980s saw a particularly rapid growth in the producer services (Table 2.2), including accountancy, computer services, consultancy, finance and market research. This attracted research to explain its causes and consequences (Box 3.2). One outcome was a challenge to the old-established distinction between goods and services production (Bailly *et al.*, 1987; Wood, 1987a; Bailly and Maillat, 1988; 1991). Most commentators, however, simply refined the sectoral view, by focusing on the growth and locational dynamics of business and financial services separately from other service activities (Stanback, 1980; Stanback *et al.*, 1981; Gillespie and Green, 1987; Marshall *et al.*, 1987; Beyers, 1989, 1990a; Leo and Phillipe, 1991).

This work developed that of Greenfield (1966) and Singelmann (1978) rather than of Bell, and especially Greenfield's insistence on the contribution to business of the combined skills and expertise of in-house white collar workers, and external business service companies. Like Gershuny, this implied that the driving force behind the most significant service growth is the tertiarization of the production process (Gershuny and Miles, 1983).

Geographical work further argued that these business services are central to the economic base of many regions. Producer services 'play a significant role in spatial differentiation because their demand and supply need not be geographically coincident and they are not dependent upon the level of economic activity in an area' (Marshall *et al.*, 1988: 14). Producer services are also important because, unlike much consumer service work, they offer full-time jobs to a highly skilled workforce (Beyers *et al.*, 1989).

Both Polese (1982) and Marshall (1982, 1983) demonstrated the scale of inter-firm exchange involving business services, and the fact that business services such as accountancy, legal services, consultancy and market research supply not only manufacturing, but also a wide range of other services, including the financial sector, government and other producer service industries. In a key study, Beyers and Alvine (1985) also show that more than half of a sample of approximately 1000 firms in the Puget Sound area of Washington State exported half their output outside the immediate local area. They conclude that,

> The linkages found in the Central Puget Sound region ... are contrary to central place theory notions that ... service firm's exports are exclusively to their surrounding periphery. Many service sectors have a base of income which is as external and spatially diverse as many sectors traditionally considered key to the local economy. (Beyers and Alvine, 1985: 42)

This accords with Noyelle's conclusion that, 'Until recently, manufacturing has been a principal component of the export base ...

Increasingly, however, it is advanced services that constitute the vital sector' (Noyelle, 1983: 286).

The dynamics of service growth, and associated patterns of uneven development, thus arise from intermediate demand for specialist services between regions, and sometimes even at the international scale (Daniels, 1986a, 1991a). These patterns are particularly moulded by the requirements of large corporate organizations. Investigations have thus also been directed to the impacts on producer services location of demand and location of market and technological change, and of corporate contracting out or 'unbundling' (Barcet et al., 1983; Beyers, 1990a; Perry, 1990a). The concentration of services close to major metropolitan regions thus reflects demands from corporate head offices, and the agglomeration advantages of networks of associated specialist suppliers of expertise. The externalization of service requirements by firms underpins this spatial centralization because,

> forces of agglomeration tend to produce what may be termed a 'complex of corporate activities': the spatial clustering and mutual symbiosis of 1) the head or divisional offices of primary, secondary and tertiary sector firms; 2) high order financial establishments; and 3) the producer service firms that provide inputs to the first two types. (Coffey and Bailly, 1992: 865)

With an increasing contracting out of services from large clients to business service firms, the advantages offered by such complexes of specialist suppliers become more pronounced.

Outside the major cities, while there may be some specialist local markets, the scope for the expansion and specialization of service functions is generally lower. Nevertheless, there remains considerable growth of business services in less urbanized locations (Wood, 1986; Kirn, 1987; Jaeger and Durrenberger, 1991). O'Farrell and Hitchens (1990b: 1992) argue that disparities in producer service provision may have a significant impact on the performance of other businesses in local economies. They show that producer service firms in peripheral areas have a narrower client base, a less qualified and skilled workforce and less wide-ranging experience (see Chapter 9).

The investigation of producer services during the 1980s thus highlighted the links between these activities, wider corporate changes, and local and regional development. They nevertheless tended to be treated in too uniform a manner, neglecting the wide variety of trends contained within them. In practice it is also difficult to separate producer services from others, especially when offered by firms also serving consumer and public sector markets (Chapter 2, page 31–3). It is also doubtful whether the distinction between consumer and business services is valid for some services, most notably the financial services. These are better viewed as being concerned with extracting profit from capital circulation, rather than imparting expertise to their business. In focusing on producer services as the key to spatial differentiation, producer service studies also neglect the combined, overlapping geographies of all services, including consumer and public services in supporting local economies (see Chapter 9).

The geography of service enterprise

There is a long-established recognition that large firms dominate the commanding heights of the economy (Prais, 1981; Dicken, 1992), and that such prime movers shape its spatial development. This is the basis of a 'geography of enterprise' approach to the spatial organization of services in the economy, drawing on the behavioural tradition in industrial geography (Box 3.2) (Watts, 1980; Wood, 1987b). Many of the characteristics of structural change, for example the breaking down of the boundaries between manufacturing and services, or the increasingly international character of some services, are interpreted as a consequence of the activities of large service conglomerates.

The corporate approach emphasizes the hierarchical character of the spatial organization of large firms, with successive layers of administration and management located in different places. The most senior, experienced and skilled staff, and those carrying out the most innovative activities, are found principally in major metropolitan areas and their hinterlands, attracted there by the size of the market, access to other corporate services, international communications, including airports, and government contacts (Wheeler, 1985, 1986, 1987; Wheeler and Dillon, 1985). Provincial sites generally offer a truncated range of more routine services, and are dependent for more advanced requirements on headquarters staff, or information provided from central offices (Glasmeier and Borchard, 1989). This constrains the local development of in-house expertise and specialist services (Britton, 1974; Crum and Gudgin, 1977; Marshall, 1979, 1983; Gudgin, et al., 1979; Daniels, 1986a; Dunning and Norman, 1987; Coffey and Polese, 1987; Enderwick, 1989; Howells, 1989).

Large business service firms work closely with large clients, but also often provide vital managerial or technical expertise for small and medium-sized enterprises. Small business service firms often spin-off from large client or service firms, and these also sometimes consciously contract out functions which they previously performed for themselves, supporting or even creating new small service companies (Howells and Green, 1986; Wood et al., 1993). The location of these small business service firms is thus shaped by the activities of large organizations (Howells, 1989). The current concentration of service functions in metropolitan centres therefore encourages the formation of new firms in these areas. In turn, the development of specialist and technical expertise encourages contracting out. Metropolitan areas, however, may be less obviously favoured by services supporting manufacturing. The reorganization of distribution, for example has encouraged the development of national distribution suppliers in less-urbanized locations (Marshall, 1989).

Much growth in service employment during the 1980s, especially in business services such as consultancy and market research, was in small firms (Keeble et al., 1991; Wood et al., 1993; Bryson et al., 1993). Indeed, much of the resurgence in the small firm sector as a whole has been led by service industries (Keeble, 1990). For a whole variety of reasons this

Box 3.2: European service studies

European service studies provide a good example of the conventional approach to service location. In the 1970s and early 1980s they documented the scale and growth of office employment (Thorngren, 1970; Daniels, 1975; Goddard, 1975) and its intense concentration in large urban areas such as the Randstad (Amsterdam, The Hague, Utrecht and Rotterdam), Stockholm, Paris, London and Dublin (Bannon, 1973; Burtenshaw et al., 1981; Bateman, 1985) More recently they have focused on 'advanced' producer or business services, deemed increasingly critical to business in an unstable and dynamic economic environment (Lambooy and Tordoir, 1985; Tordoir, 1991). Such higher-order professional services are growing rapidly in capital cities but are also decentralizing to less-urbanized locations (Hessels, 1992). Since the late 1980s the contribution of a range of services to regional development has been considered (Illeris, 1989b; Daniels et al., 1993).

Though the interests of European studies have broadened (De Smidt, 1984, 1985, 1990), many studies stress the impact of 'agents of change' i.e. business enterprises and property institutions on service location. Increasingly these are analysed in an international comparative context (Daniels et al., 1988, 1992). Building on neo-classical approaches to industrial location (Weber, 1929; Daniels, 1985; Smith 1987; Chapter 4), early office research derived an interest in the role of linkages between firms in determining their location (Goddard, 1973, 1975). But following Pred (1967) research broke with classical theory on the grounds that it failed to take account of the impact on service location of the diverse perceptions and motivations of managers in multisite companies.

Thus, this work displays well both the strengths and weaknesses of a behavioural or corporate approach to service location. It examines in intricate detail the character of individual offices and the companies they are in, how these may differ between individual places, and the way differences in the locational preferences of firms influence location (Edwards, 1983; Van Dinteren, 1987). Business service firms are depicted as a spatially dynamic sector leading service growth in metropolitan areas. Their need for proximity and accessibility to clients and other service firms is seen as the key to their spatial concentration. However, tendencies towards centralization may be undermined by the cost of accommodation, lack of room for expansion and congestion in the city centre, which encourages some firms to decentralize (Hessels, 1992).

The analysis says little about the character of employment provided by different services, or the local milieu and competitive context in which they work. With the notable exception of work by Bailly (Bailly and Maillat, 1988, 1991; Bailly et al., 1987), business services are largely treated as a 'stand-alone' sector; there is little investigation, over and above communications patterns, of their relationships with their business, public sector or consumer clients. We thus learn little about the markets firms serve, what is necessary for commercial success in these markets, and what it is about them that makes it important to be accessible to and to have good communications with customers.

trend has tended to widen spatial disparities in service employment growth. For example, the highly qualified personnel most likely to set up small service firms are over-represented in prosperous areas (though not necessarily the large urban centres), and these areas are also more able to sustain larger numbers of small firms serving local markets.

Conclusion

The approaches discussed in this chapter directly address the overwhelming evidence for the scale and growth of service activity in modern advanced economies. They have generally sought to identify the key economic and social mechanisms that explain these phenomena. Different commentators have given priority in such explanations to different characteristics of modern society. For some the expanding power of consumer choice is particularly important, while others emphasize the growing labour productivity gap between the manufacturing and service sectors. Some have associated service growth with the impacts of new information processing and communications technology, as part of the increasing economic preoccupation with information, rather than material activities. Others point to the developing complexity of all production processes, and its effects on patterns of interrelationship between sectors, and between production and consumption. Service functions have undoubtedly expanded in response to all of these and other changes.

Changing service location patterns have also been explained variously, as a consequence of developing information-processing networks, the shifting attractions for office functions of major metropolitan areas, the increasing spatial segregation of economic functions affected by modern communications technology, and the impacts of corporate restructuring, favouring the growth of specialist service sectors. Again each of these, with other pressures, support the increasingly uneven geographical distribution of much service provision. Essentially, services are now exchanged over longer distances, while the control of critical information and expertise is centralized.

These forms of explanation tend to be impressed by the leading role of at least some service activities in modern economic and geographical change. This is because the elements of service change which they chose to examine are particularly dynamic. The approaches to which we shall turn in Chapter 4, however, tend to place this dynamism in a wider context of capitalist development, to the point of arguing that service change is a by-product of other, more fundamental, trends. At the end of the next chapter, we shall elaborate our own perspective which draws on both traditions.

Further discussion and study

Guided reading

Bell D 1973 *The coming of post-industrial society*. Heinemann, London

Gershuny J 1978 *After industrial society? The emerging self-service economy*. Macmillan, London: Chapters 5–7.

Gottmann J 1983 *The coming of the transactional city*. Institute for Urban Studies, University of Maryland, College Park, MD

Hepworth M 1989 *Geography of the information economy*. Belhaven Press, London

Activity

1. During any week, make a diary or sources of *information* you employ, including newspapers, books, radio, TV, word of mouth, public meetings, advertisements, or any others. Note where such sources originate, and how the information was delivered to you.

    What proportion depends on the use of manufactured goods, and upon modern information technology?

    How much of this information made a difference to your behaviour immediately, or might do so at some time in the future?

    Compare your experience with the experience of others. Do all groups in society have equal access to sources of information? If there are any differences what implications might this have?

    How might the mix of sources of information have been different in 1960, 1930 or 1900?

    What do you conclude about the impact of the 'information revolution on modern life'?

2. Examine any modern office as an 'information processing' unit.

    (a) *What is the service 'product' it offers?* What inputs/outputs of paper and information are required; where do directives for action originate, and who are the office's 'clients'?

    (b) *How is work structured in and around the office, and why?* Does it have an open/divided layout; what is the age, gender or status of workers; what skills are required; who initiates and supervises work; what decisions are made routinely, regularly, or are required unpredictably; who cleans and maintains the building?

    (c) *What connections does the office have with others parts of its own organization or other organizations?* To what extent is this by post, telephone, computer or video link? How has it changed in recent years? What implications have any changes had for the operation and control of the office?

    (d) *What impact has information technology had on work?* How have the following changed: filing; communications; typing; data analysis; dependence on computer software; supervision? What impact has this had on the people who work there?

Consider

1. What is meant by 'high level', as opposed to more routine service activities, and which commercial and public services are primarily involved in each?

2. Take a number of typical business services, e.g.: computer services,

management consultancy, training consultants, research and development, accountancy, legal services, marketing and selling, and design.

What sources of demand can you identify for these services? To what extent are these services supplied by companies for themselves or by specialist suppliers? How might demand for these services be changing and why? How might demand change in the future?

Discuss

1. 'Modern communications technology should encourage the geographical dispersion of all but the highest-level services.' How far is this a valid summary of current spatial trends in different services?
2. How far is the concentration of office complexes in cities a consequence of their need to communicate together?
3. To what extent does the growing prominence of services imply the development of an information economy?

CHAPTER 4

'Alternative' approaches to service growth and location

The 'conventional' approaches to service growth and location evidently offer wide and overlapping variations of interpretation. Indeed, this lack of theoretical coherence offers perhaps the most legitimate basis for criticizing them. An 'alternative' approach has emerged which specifically attempts to present a coherent and critical perspective, based largely on a Marxist view of the general progress of 'late capitalism' or 'late industrial' society. This builds upon nineteenth century political economy, and especially emphasizes continuity rather than any significant change in the evolution of capitalist societies. Such critical responses to conventional interpretations of service growth gathered pace during the 1970s and 1980s. They challenged the whole notion of the new 'service economy' as a significant break with past trends in industrial capitalism. While the growing significance of service functions is accepted as an important structural economic change, it is seen as reflecting the same processes which created earlier patterns of manufacturing growth; the search by capital for new forms of production as old forms are undermined by competition (Mandel, 1975; Allen, 1988a). Despite service employment growth, the essential basis of wealth creation remains in the material processing, extractive and manufacturing sectors.

The acknowledgement of the significance of service growth at least marks a change from past Marxist preoccupations with manufacturing processes and blue collar work, and to an extent with services as 'non-productive' activities (Sayer, 1985; Allen, 1988b, 1992; Winckler, 1990; Sayer and Walker, 1992). Service growth is now one facet of the search for profitable investment, as returns on mass production decline. The relationship of service developments to material production, however, is still emphasized. This is thus a production-based view of change, rather than the market- or demand-based perspective characteristic of 'conventional' approaches (Allen, 1988c).

Scott makes explicit the difference represented by this perspective compared with 'conventional' analysis (Scott, 1988a). For him, Bell and his derivatives associate the growing prominence of services too readily with genuinely new forms of capitalist development:

> The postindustrial hypothesis ... [is] utterly wrong insofar as it points to the latent transcendence of capitalism by a sort of new information-processing mode of economic organisation. Of course, we must acknowledge that contemporary capitalism is in part distinguishable from earlier forms by its greatly expanded dependence on white-collar workers; its burgeoning business and personal service functions, the massive increase in banking and financial operations. ... To mistake them, however, for signs of a fundamental shift away from industrial capitalism is to fail to understand their role and purpose in modern society'. (Scott, 1988a: 7)

Sayer and Walker also take issue with 'conventional' theorists who argue for the rise of an information economy:

> Some service theorists think we have entered an age of information and communication. ... The information explosion in the contemporary economy is readily apparent, but information is not a free-floating ether; it must be pinned down. Information can be either part of industrial products or ... part of their production, circulation, or consumption. ... The electronic age has unquestionably increased the ability to package greater amounts of information in more sophisticated products, and industrial output has become more information-rich, but there is no convincing evidence for a large, distinct information-producing sector in the economy. (Sayer and Walker, 1992: 63)

For them, what is important is not that industrial activity has become more intangible or 'informational', but that the growth of office-based services is part and parcel of the way in which capitalism creates new means of both producing and distributing surplus value (that is the difference between the cost of labour and the market price of goods).

The influence of Marx

The preoccupation with material production in Marxist-based analysis, and the interpretation of changes in services as a response to adjustments in manufacturing, can be traced back to Karl Marx. Marx identified the pursuit of surplus value through investment of capital as the core motive driving capitalism. This surplus was achieved through the exploitation of labour, and ultimately realized through processes of market exchange. Marx did not regard the distinction between manufacturing and services as significant, but services became implicated in a debate, going back at least to Adam Smith, over the difference between 'productive' and 'unproductive' labour. For Marx, 'productive' labour added to the sum of society's wealth, in that its product both possessed a use value and contributed to surplus value. 'Unproductive' labour, in contrast, is financed out of revenue or profit. Such labour may support circulation, capital hoarding or consumption, but it does not add to aggregate wealth (Gough, 1979).

Though Marx rejected Adam Smith's association of 'productive' labour solely with material production, quoted in Gough, 1972: 48–52), he believed that, in practice, because the capitalist mode of production had only penetrated commodity production, 'productive' labour was largely confined to those directly involved in material transformation (quoted in Gough, 1972: 52–3). Marx defines this quite broadly to include 'those who contribute in one way or another to the production of the commodity, from the actual operative to the manager or engineer (as distinct from the capitalist)' (quoted in Gough, 1972: 56).

Nowadays this would include large numbers of technical, managerial and administrative staff supporting material production. But such a definition would need to be adjusted to accommodate the mushrooming of a vast array of professional business service firms, less directly supporting material production. Marx also regarded much transport and distribution as branches of industrial production, since they were required to bring commodities to the final consumer. But public services financed out of revenue, service workers involved in circulation such as buyers, sellers and advertisers, and domestic and other personal services, which add nothing to surplus value, were regarded as 'unproductive' and therefore not central to the formation of class relations. It is not surprising, then, that large parts of the modern service sector for many years received little attention in Marxist writings.

Mandel suggests that Marx was inconsistent in defining 'productive' and 'unproductive' labour (Mandel, 1975: 401–7). He also recognized that 'unproductive' capital may indirectly increase surplus value, for example, by reducing the turnover time of productive capital, or providing credit. Once the indirect relations between economic functions are pursued, the 'productive–unproductive' distinction is therefore difficult to sustain (Brewer, 1984).

Marx also envisaged a specialization of functions into an increasing division of labour as the economy grew. Specialization of labour functions occurs both within firms (the detail division of labour) and between them (the social division of labour), and reflects capital's attempts to control labour and increase the efficiency of its use. In one of the most influential recent Marxist-derived commentaries on the growth of services, Walker develops the notion of a changing and extending division of labour, to provide a sophisticated analysis of the emerging complexity of service activity (Walker, 1985; Box 4.1). The distinction between 'productive' and 'unproductive' labour is replaced by that between 'direct' and 'indirect' labour, with many activities treated as services in the 'conventional' literature, described as indirect labour in relation to the material basis of production.

Walker recognizes that capital has been directed into more indirect service-intensive forms of material production, including new areas of knowledge-based production and an increasing variety of tradable consumer services, such as global tourism and leisure. The financial services have also grown rapidly to support and augment the international search for profitable capital investment. Business services have increased as managerial, professional and technical skills have become increasingly separated from routine factory or office work.

57

Box 4.1: Walker's view of services

Walker is critical of exponents of a post-industrial or service-led economy. Instead he gives priority to the production of goods in the economy, and sees services as more or less directly concerned with their manufacture. In his view, much of the growth of services reflects the increasingly indirect nature of labour in support of goods production.

His first line of argument is that the growth in services is often exaggerated because conventional approaches adopt a cavalier attitude to the definition and classification of services, and incorporate many activities which are more readily defined as goods (Chapter 2, pages 33–6). He suggests a distinction between goods and services based on both the tangible form of the output and the type of labour involved. Thus,

> The distinction between goods and services lies in the form of labour and its product. A good is a material object produced by human labour. . . . A labour-service, on the other hand, is labour that does not take the intervening form of a material product. . . . It is thus normally irreproducible by other workers and involves a unique transaction between producer and consumer. (Walker, 1985: 48)

This defines much conventionally recognized service growth as aspects of goods production including administrative and managerial support to manufacturing, and service industries associated with physical output. 'Services' are confined to a small group of personal services including the performing arts, professional advice and waitressing (see Chapter 2).

The second element of his argument is that much of the service growth charted by 'conventional analyses' reflects a widening and deepening of the division of labour associated with material production, in response to the growing sophistication, complexity, scale and geographical dispersal of manufacturing. Service growth is largely in what he terms 'indirect labour' related to material production; in white collar staff working in-house in manufacturing, and externally supplied business services. Contrary to conventional interpretation, this is not evidence for a service-led economy. The close integration of goods and 'service' production denies the possibility of any independent service dynamic.

On the other hand, Walker does not produce a simple restatement of the primary role of manufacturing in the economy (Allen, 1992). He acknowledges that the 'locus of competitive advantage' has shifted in goods production from simple production efficiency to indirect activities, including advertising and marketing, technical, research and distribution activities, conventionally defined as services (see pages 29–33). Personal consumption is also regarded as an input to production, for example providing the education and skills necessary for a successful labour force. Consumer service industries also provide the necessary means to bring goods to the consumer. He notes, like Gershuny, that there is little evidence for a shift in consumer purchases in favour of services rather than goods (Walker, 1985: 60). This reinforces Walker's third argument that changes in production rather than consumption are the explanation for 'service' growth.

State-provided services in infrastructure, educational and social services, in trade and other regulation, and in defence also, in his view, have expanded to meet the needs of capitalist production.

Services have therefore contributed powerfully to economic and social change. From Walker's perspective, even local consumption can be regarded as an input to production, serving the profit-making requirements of capital, sustaining the reproduction of labour, and acting as the means by which capitalism can penetrate and dominate the domestic household. He thus concludes that, today, the realm of 'indirect' (or service) labour, 'may be the principal locus of industrialism, to which most product development . . . is directed' (Walker, 1985: 80).

Walker's emphasis on services as 'indirect' labour retains the Marxist priority given to material production. Instead of excluding services from production, however, he argues that activities organized to support the interests of capital, even if only indirectly, must serve some function in relation to the creation of surplus value, and are thus essentially productive.

Arguments about the productive role of different economic functions are perhaps less important than scenarios of change derived from 'alternative' perspectives. There are some obvious sources of instability in the shift to services. It offers increasingly polarized work opportunities, between the rich and poor. Services jobs are also as often subject to displacement by investment in manufactured substitutes as any other work. Mandel (1975) saw the crisis of late capitalism as a chronic shortage of opportunities to employ a growing surplus of capital in material production. Since the mid twentieth century, global competition and levels of investment risk have deterred manufacturing investment in many countries. Capital has instead been directed to more profitable, short-term information exchange, financial speculation and consumption activities. Service growth is thus fundamentally unstable: 'A society consisting only of service trades, in which the entire proletariat had become unproductive wage-earners could not be sustained as capitalism' (Mandel, 1975: 407). Mandel's position thus appears in some ways to be similar to that of 'conventional' pessimists concerned about the negative impact of service growth.

Service growth and location

The restructuring approach

As among 'conventional' approaches, consideration of the locational implications of service change reveals differences of emphasis within the 'alternative' tradition. Massey (1979 a,b, 1984) advanced the principle of the spatial division of labour, which affects the number and characteristics of jobs available in different places. The 'division of labour' refers to the pattern of work specialization in production, developed with increasing intricacy over time to ensure the efficient use of capital investment (see Chapter 5). This pattern includes a fundamental spatial component, expressed in the UK during much of the twentieth century, in the specialization of distinctive regions into particular products and trades, such as coal mining, metal work, textiles or port work. These supported dominant local forms of capital–worker relationships, labour skills, and social and community patterns, including traditional roles for male and female workers.

In Massey's early work, modern change away from these patterns of regional industrial specialization primarily reflects the growing dominance of large manufacturing firms. While pursuing the same profit goals as of old, these can now deploy resources on an inter-regional, and even international scale, seeking out new opportunities to exploit labour in different places. Not only can they transfer production work to new, cheaper locations, but they also tend to separate various types of white collar work, including control and research functions, from blue collar production. A new spatial division of labour has become dominant, therefore, in which regional functional differences have emerged since the 1960s, with some dominated by 'control' activities (often classified as 'services') and others confined to more routine production.

The processes of capitalist industrialization have not changed, therefore, but the form, including the spatial form adopted, has been transformed. The patterns of change are nevertheless influenced by the earlier patterns of work, including regional specialization. The legacy of past labour practices, and of past geography, have affected how regions have fared in the new 'round' of investment. Many old industrial and mining regions have attracted investment only in poorly paid routine production, often mainly for women, to replace their traditional male-dominated, unionized activities. They thus increasingly depend on state subsidies and service support. Formerly unindustrialized, rural and small town regions, with little legacy of industrial practices and their associated social attitudes, have drawn in more new investment. Meanwhile, the higher-income, growing control functions, with associated business and consumer services, have focused in traditionally service-based metropolitan regions, in the UK dominated by the South East of England.

The changing 'spatial division of labour' has been described as 'less a theory, more a heuristically valuable metaphor' (Warde, 1985: 195), which suggests that the nature of relationships between historical inheritance, spatial structure and modern economic and social change are complex (Lovering, 1989). Nevertheless it explains, if only in outline, the emergence and regional concentration of business services, and research and development functions, in terms of the inherited social characteristics and labour relations of different regions (Massey, 1984).

The restructuring approach has also been extended to private consumer and public sector services. In an analysis of the mechanisms causing the decline of manufacturing employment, Massey and Meegan (1982) highlighted three forms of production reorganization affecting levels of employment: work intensification, rationalization of production, and investment and technical change. Pinch (1989) augmented the work of Urry (1987) to show how these categories are also applicable to public as well as private services (Table 4.1). In private sector consumer services, it might be expected that there would be the same profit-oriented pressures towards greater efficiency as in manufacturing, for example through increased labour productivity, closure of sites and concentration into larger units, and changes in the role and training of labour. Pinch demonstrates, however, that similar outcomes have arisen from the 'commercial' criteria of efficiency now being sought for public services (see reading, Chapter 8, pp. 197–204). This restructuring involves the 'commodification' of some services, through their replacement by manufactured technology (e.g. investment in technical change, 'materialization' – see Table 4.1). There are also significant

Table 4.1 Forms of service sector restructuring

Private sector	Public sector
1. Partial self-provisioning	
Self-service in retailing; replacement of services with goods; videos, microwave ovens, etc.	Child care in the home. Care of elderly in the home. Personal forms of transport. Household crime prevention strategies, neighbourhood watch, use of antitheft devices, vigilante patrols.
2. Intensification: increases in labour productivity via managerial or organizational changes with little or no investment or major loss of capacity	
Pressure for increased turnover per employee in retailing.	The drive for efficiency in the health service. Competitive tendering over direct labour operations, housing maintenance, refuse collection. Increased numbers of graduates per academic in universities.
3. Investment and technical change: capital investment into new forms of production often with considerable job loss	
The development of the electronic office in private managerial and producer services.	Computerization of health and welfare service records. Electronic diagnostic equipment in health care. Distance learning systems through telecommunications video and computers. Larger refuse disposal vehicles, more efficient compressed loaders.
4. Rationalization: closure of capacity with little or no new investment or new technology	
Closure of cinemas.	Closure of schools, hospitals, day-care centres for under fives, etc. Closure or reduction of public transport systems.

Table 4.1 Forms of service sector restructuring *continued*

Private sector	Public sector
5. Subcontracting: of parts of the services sector to specialized companies, especially of producer services	
Growth of private managerial producer services.	Privatization or contracting out of cleaning, laundry, and catering within the health service.
6. Replacement of existing labour input by part-time, female or non-white labour	
Growth of part-time female labour in retailing.	Domination of women in teaching profession? Increased use of part-time teachers.
7. Enhancement of quality through increased labour input, better skills, increased training.	
In some parts of private consumer services.	Retraining of British Rail personnel. Community policing?
8. Materialization of the service functions so that the service takes the form of a material product that can be bought, sold and transported	
Entertainment via videos and televisions rather than 'live' cinema or sport.	Pharmaceuticals rather than counselling and therapy?
9. Spatial relocation	
Movement of offices from London into areas with cheaper rents.	Relocation from larger psychiatric hospitals into decentralized community-based hostels. Relocation of offices from London to realize site values and to reduce rents and labour costs.
10. Domestication: the partial relocation of the provision of the functions within forms of household or family labour	
Closure of laundries.	Care of the very young and elderly in private houses after reductions in voluntary and public service.
11. Centralization: the spatial centralization of services in larger units and the closure or reduction of the number of smaller units	
Concentration of retailing into larger units. Closure of corner shops.	Concentration of primary and secondary hospital care into larger units, that is, the growth of large general hospitals and group general practices.

Source: Pinch (1989)

domestic and social implications, as with industrial restructuring, through increased 'self-provisioning', rationalization and domestication.

Other applications of the restructuring approach to services emphasize the differing and overlapping geographies of a variety of services, including public compared with private services, internationally tradable services compared with those serving domestic markets, and financial and commercial services compared with consumer and tourist services (Urry, 1987, 1990a,b; Allen, 1988a). The main focus, however, is still on the type of employment offered by service industries (Buck, 1985). One much-discussed theme is the rise of a 'service class' of professional and technical workers as a significant intervening stratum in society between capital and labour (Abercrombie and Urry, 1983; Urry, 1986; Thrift and Leyshon, 1992; Savage et al., 1992). Restructuring analysis also emphasizes the polarized nature of service employment growth, with the parallel expansion of casual and less-skilled service work, especially in cleaning, catering, security and retailing services (Allen, 1988b, 1992; Little et al., 1988; Christopherson, 1989; Christopherson and Noyelle, 1992; Crang and Martin, 1993; Allen and Henry, 1994).

Gender issues have also been introduced into the analysis of service location. The supply of female labour, for example, is seen as a key factor influencing the relocation of offices from central business districts (McDowell, 1983; Nelson, 1986). Another focus is the complex interaction of women's domestic and formal work responsibilities, and inequalities in the provision of state support services (McDowell, 1992). Such issues are taken up more fully in feminist analysis of urban life which go well beyond the boundaries of the restructuring approach (Pratt, 1990; Bondi, 1990; McDowell, 1993 see Chapter 5). This emphasizes the patriarchal social context of much service work, underlying markedly different trends in male and female employment (McDowell, 1991; Massey, 1991).

Recent research also demonstrates the increasing significance of service employers in different localities (Cooke, 1989), and similar interrelations between past and present, and local, national and international patterns of service investment, to those suggested for manufacturing (Allen, 1992). Urry (1990b,c,d), though, recognizes the potential even of a range of consumer services (the arts, leisure, recreation and tourism) as sources of regional economic growth. He indicates that

> there are prima-facie grounds for thinking that in some countries or regions the characteristic relationship of a manufacturing base generating a certain quantity of other activities has been reversed so that it is service industry which can now be viewed as the base. (Urry, 1990d: 1–2; quoted in Allen, 1992: 295)

This conclusion, of course, is quite different from the traditional Marxist priority given to blue collar production work. It parallels interpretations in the 'conventional' tradition, based on the analysis of business service exports, and the contribution of such services to manufacturing competitiveness.

Phases of capitalist accumulation

The restructuring approach focuses on the interaction of production and place in creating changing spatial patterns. Other perspectives show how different phases in the wider logic of capitalist development produce dominant patterns of location. Central to the ideas of Harvey, for example, is the tension between the inertia of fixed investment in particular places, necessary for profitable production, and the constant flux of capitalist investment, seeking out new opportunities for the extraction of surplus value. Harvey argues that capitalistic pressure to reduce the time between investments, to ensure a flow of profitable returns, requires the reduction of spatial barriers in the economy. As new profit-making opportunities arise, this inevitably opens up 'fresh geographical spaces for accumulation' (Harvey, 1985: 44), thus hastening the destruction of investment in existing places.

Harvey and other commentators have analysed this tension between fixity and crisis, drawing to varying degrees on French regulation theory. Phases of development shape and transform geography not just through the organization of production, but also through characteristic forms of consumption, and patterns of state regulation. The post-1945 era in the industrialized countries brought to fruition a 'Fordist' regime of accumulation, based on the large-scale production of standardized consumer goods, and a highly specialized division of labour within production. This form of organization was also sustained by 'corporatist' national agreements over wages and working conditions between management and trades unions. Government demand management, on Keynesian principles, and welfare state spending underpinned the mass market. Regional policies, promoting industrial decentralization, supported demand in areas with high levels of unemployment, supposedly relieving inflationary pressures in more prosperous regions. Financial stability was ensured by an internationally regulated system of monetary coordination (Boyer, 1986; Lipietz, 1986; Aglietta, 1979).

This system is regarded as having reached the limits of its development during the 1970s. It was undermined by the cost and slowness of technical change in production processes, especially as international competition intensified from countries with lower labour costs or higher investment in newer production technologies. Capitalism thus sought an alternative 'regime of accumulation' which would revive profitability. This involved computer-coordinated technological innovation enabling new forms of production to respond more readily to different market demands. New ways of organizing labour were also progressively introduced, at least into key sectors, to improve productivity, and avoid the conflicts inherent in the Fordist system (Gertler, 1988, 1992; Schoenberger, 1989). Cutbacks were also required in public services and welfare provision, and there was a general reduction in the economic role of the non-market, public sector. A more entrepreneurial approach to governance, often involving the private sector, was imposed to reduce costs (supply side adjustments), rather than stimulating the expansion of demand (Jessop *et al.*, 1991).

Harvey regards transportation and communications innovations as central to the development of this new phase of accumulation (Harvey, 1989a). They reduce the turnover time of capital by speeding up the pace of life and reducing the time horizons of decision-making. Economic activity is thus integrated over a wider, even global space, releasing some of the constraints created by the fixity of investment and allowing greater responsiveness to new demands. The role of services, whether enhancing circulation, serving more varied consumption needs, or involving a changing role for the state, would thus appear to be central to this interpretation of economic and spatial change. Yet some who draw on the regulationist approach still cling to the traditional Marxist view that, 'service industries are essentially "parasitic" in that they do not actually add wealth in the economy, although they can help to realise the value of the wealth created elsewhere' (Peck and Tickell, 1991: 360).

More generally, manufacturing sectors such as motor vehicles, clothing or electronics are seen as the core of the economy, setting its tempo and acting as harbingers of change (Storper and Walker, 1989). One view, for example, is that changes associated with microelectronics are leading to major transformations of the industrial landscape as new groups of industrial producers seek out locations uncontaminated by traditional industries, on the margins of old industrial regions, or in less urbanized environments. Elsewhere, horizontal and vertical disintegration of control to smaller production units and firms, associated with the search for greater flexibility of operations to serve changeable markets, encourages the development of more self-contained regional industrial complexes (Scott and Storper, 1986; Scott, 1988a,b,c).

While some reference is made to business service agglomerations in metropolitan regions, services are largely neglected. Scott, for example, devotes only four pages to the development of office industries in his book on metropolitan development (Scott, 1988b). Yet the manufacturing industries described in this literature are heavily tertiarized, and the evidence presented of corporate restructuring involves a significant non-production component (Shutt and Whittington, 1987). The focus on manufacturing also belies the fact that supposedly new, more flexible operations of modern manufacturing have for a long time been characteristic of services; for example, flexible management styles, part-time employment, small firms, vertical disintegration, intense local interaction and heavy investment in information technology (Pollert, 1988). It is tempting to suggest that the emphasis on disintegration and local agglomerations in this work arises precisely because of the growing tertiarization of the manufacturing production, as it becomes more like service activity.

Even where the growing prominence of services is analysed, as in the Storper and Christopherson reading in Chapter 6, they are treated very much like manufacturing production. The core role of the manufacturing sector is also still maintained. This is because many believe that service employment growth is unlikely to ameliorate fully the effects of deindustrialization in the advanced economies (Petit, 1986). Services will not

> lead to a lasting recovery ... [or] the long term establishment of a stable model of slow growth. ... The current period seems to be a transitional phase [in which] the predominant Fordist model of industrial growth of the past has, at the level of individual economies, revealed its limits. (Petit, 1986: 195–6)

The evident growth of business services is nevertheless acknowledged, and it is suggested that, 'the division of labour entered a new era in the 1970s and 1980s: firms started to farm out a wide range of tertiary activities [and] [i]nstead of seeing services and manufacturing as different ... they should be seen as interdependent and interrelated activities' (Coriat and Petit, 1991: 19). As with Walker's analysis, this deepening relationship between services and their manufacturing clients is seen as a form of (re)integration of services within the manufacturing sector. It is not part of a transition to a service economy but part of a general process of industrial evolution, widening and deepening the social and technical divisions of labour within manufacturing (Sayer and Walker, 1992).

'The power of money'

In parallel to these manufacturing-oriented interpretations, an interesting branch of research has paid specific attention to the 'power of money', and the role of the increasingly autonomous financial sector in the crisis of Fordism (Thrift, 1994; Leyshon, et al., 1988; Leyshon and Thrift, 1992, 1993). In the last two decades the 'financial system has moved away from its role as a facilitator of the production and distribution of goods ... it has taken on a life of its own, a fact that can be seen vividly in the mushrooming of speculative activity' (Magdoff and Sweezy, 1987: 21). From this perspective the limitations of the post-1945 phase of growth are less to do with the break up of mass markets or technical rigidities in the production process, but rather 'the growing hegemony of finance capital over productive capital ... as finance capital began to take advantage of the break down of the international regulatory structure ... which [was] set up to control finance capital in the interest of national processes of capitalist accumulation' (Leyshon and Thrift, 1990: 6–7).

The decision to move to floating exchange rates led to the break up of the Bretton Woods agreement in the early 1970s, and the subsequent development of a new international financial system outside national control (Thrift 1987; Thrift and Leyshon, 1988). Thrift (1990a) shows how the 'deregulated' financial sector has transformed spatial development, especially driving growth and wealth creation in major metropolitan cities. Although ultimately related to economic activities based elsewhere, this only partially trickles down the urban hierarchy

(Leyshon and Thrift, 1989). Change and volatility in the financial sector also penetrate other sectors shaping metropolitan centres. Real estate developers, for example, increasingly employ novel forms of finance capital which give

> considerable fluidity to the way developers . . . can conceive of and roam space . . . thereby promoting activities on land that conform to the highest and best commercial uses . . . and fostering an urban landscape extremely sensitive to the temporal aspects of accumulation as dictated by interest rate fluctuations and supply and demand for money capital. (Merrifield, 1993: 111)

The development of metropolitan centres is thus increasingly shaped by global financial sector growth and speculation.

Harvey (1989a) also suggests that innovations in financial services play a central role in the advanced economies. Writing before the recession at the end of the 1980s, he relates the growing dominance of short-term perspectives in the UK and US economies, and increases in 'paper entrepreneurialism', to the dominance of the financial sector. Financial turbulence and uncertainty in these countries create instability elsewhere, and the conditions leading to the corporate search for enhanced flexibility described by Scott and Storper (1986).

Volatility in location is paralleled by increasing competition between nations, regions and cities to attract mobile investment (Harvey, 1989b). Services related to consumption, such as leisure, recreation, culture and the arts, play an important part in this competition because of their role in shaping the image of places and, ultimately, the expectations which govern investment. The growing prominence of the 'instantaneous event' or the spectacle is also encouraged in the city; in other words aspects of consumption. By these means the turnover time of capital can be further reduced, increasing the scope for profit-making.

Services have thus attracted belated attention from structuralist commentators (Sayer, 1985; Urry 1990a,b,c,d; Allen, 1992). This has occurred partly as a consequence of a wider debate over the contemporary course of 'post-Fordist' capitalist accumulation (Petit, 1986). The global and autonomous nature of the financial services, the way financial speculation shapes urban development and leads to new forms of consumption, and corporate restructuring to satisfy the short-term needs of finance capital have also been particularly influential in this shift of interest (Harvey, 1989a).

Such 'alternative' approaches to understanding the role of services in economic change, undoubtedly provide a more comprehensive perspective than the more 'conventional' discussions of service growth reviewed in Chapter 3 (Peet and Thrift, 1989). The emphasis on the widening division of labour is important. Also, the significance of finance capital, and of the changing relations between state and private sector activity, strikes positive chords for anyone familiar with developments in the US or the UK during the 1980s. But the emphasis on the dynamics of manufacturing remains questionable. It harks back to the old division between productive and unproductive labour. Even in Walker's analysis, it may be questioned whether he has yet sufficiently acknowledged the importance of service expertise, both inside and outside the financial sector, as a dynamo creating the regime of accumulation dominating the advanced economies into the twenty-first century. Despite such criticisms, perhaps the most impressive lesson to be derived from Marxist perspectives is that services functions are implicated in the powerful mechanisms by which capitalism has adjusted its use of labour, in response to increasingly complex technological needs, and the expanding world scale of its activity.

Contrasting 'conventional' and 'alternative' perspectives

Geographical work on the service-related processes underlying urban and regional change, has drawn on both economic and sociological perspectives. Inevitably 'conventional' and 'alternative' traditions have frequently overlapped. 'Conventional' studies demonstrating the contribution of services to employment growth have encouraged other writers to examine the issues. Gershuny's (1978) ideas were widely reviewed and appreciated in the early 1980s by Walker and Urry. A consensus also seems to have emerged over the significance of the financial services for economic globalization, international trade and regional specialization (Marshall *et al.*, 1988; Harvey, 1989a; Warf, 1989, 1990, 1991). In recent years, 'conventional' writers, like their counterparts in the 'alternative' tradition, have also become more pessimistic concerning the impacts of service growth (Noyelle, 1984).

Despite these similarities, 'conventional' and 'alternative' approaches to service location and change do give a different emphasis to the following:

1. *The distinctiveness of service developments.* 'Conventional' approaches tend to develop special theories of the role of the service sector in the space economy, often in terms of some aspect of a distinctive information-based 'service economy', rather than analysing service location as part of broader structural economic change.
2. *The classification of services.* 'Conventional' approaches accept sectoral classifications of services, although sometimes reclassifying official statistics to highlight the special characteristics of services, such as their producer or consumer orientation, information-intensiveness, relative tangibility or immediacy. 'Alternative' approaches challenge the separation of manufacturing and service sectors. Many activities classified as 'services' in the 'conventional' approach are treated alongside manufacturing as part of production. 'True' services are defined much more narrowly, and stress is placed on their social context, especially within a wider division of labour.
3. *The production processes of services.* The significance of the profit-seeking behaviour of industrial capital in controlling service

production is given much more prominence in the 'alternative' literature.

4. *Patterns of demand for services.* The 'conventional' approach emphasizes more the growing volume and diversity of demands for services and the way these shape their growth and change.

5. *The degree of continuity or change implied by the growth of services.* Paradoxically, perhaps, the 'post-industrial' vision of 'conventional' analysis envisages a more radical break with the past than the 'alternative' perspective, which sees service growth as merely part of a redirection of industrial capitalism.

6. *The implications of service growth for the quality of employment.* While both the 'alternative' and 'conventional' traditions now recognize the significance of the growth of service employment, the implications drawn for its wider significance vary. Optimistic 'conventional' interpretations emphasize high-quality work, replacing routine manual and clerical labour. More pessimistic commentators are sceptical about the economic sustainability of such a situation. The 'alternative' tradition generally gives greater emphasis to the variety of employment impacts, and the resulting distribution of income and wealth, including their gender and class implications.

7. *The sources of service growth.* Varying weight is given to whether the processes of service production and delivery follow or lead other changes. 'Conventional' approaches tend to see services themselves as providing an important engine of growth, performing a leading role in the economy, for good or ill. 'Alternative' approaches more readily play down the significance of services for long-term economic development, or link service changes to adjustments in manufacturing production.

8. *The significance of services for other activities.* 'Conventional' analysis emphasizes the significance of service inputs for the performance of other activities, organizing the circulation of information, ideas, people and capital, to support, production, exchange or consumption. In general, 'alternative' approaches interpret the role of services in such relationships as much more dependent on other events.

9. *The significance of technological change for service growth.* 'Conventional' approaches particularly emphasize the importance of communications and data processing innovation for service growth, often seeing them as leading service developments. The 'alternative' perspective views such innovations as no more than offering new opportunities for capital restructuring which may be exploited by investment in service functions, especially by offering a faster turnover of returns on capital.

10. *The role of public services.* The role of the state in supporting economic restructuring and, through its services and regulatory regime, moulding service development, is more prominent in the 'alternative tradition'. 'Conventional' approaches tend to depict public sector services as a drain on the productive sector of the

economy, even though private sector functions depend on state provision for many basic services and markets.

A service-informed view of economic restructuring

This and the last chapter have reviewed others' approaches to service location. In this final section, we argue for what we term, a 'service-informed' perspective on service growth and change, the character of which is developed in subsequent chapters.

Most 'conventional' approaches, as we have defined them, focus on particular aspects of service development, and generalize from these too readily to the service sector as a whole. In addition, they neglect broader changes in the industrial economies, or often see them as driven by service changes. The 'alternative' approaches, with which we are in greater sympathy, present a more integrated treatment of the diverse character of producer, consumer and circulation services, and their interrelationships with each other and other functions, including manufacturing. Some of the wider consequences of service development are also more fully addressed, especially the polarized character of service employment. In general, however, they still assign a limited role to service functions in economic change, including growing geographical inequality, though these criticisms do not apply equally to all commentators and all service activities.

Our response to these weaknesses is to seek to develop a 'service-informed' view of economic restructuring and its geographical consequences (Marshall and Wood, 1992). Unlike parts of the 'conventional' tradition, this does not involve any special theory of the role of all services in structural change. The diverse nature of services and their deep relations with other forms of economic activity argue against this. Rather, we agree with the 'alternative' approach that service developments must be understood as a component of wider processes of economic and social restructuring, which are shaped by the demands of profitable production in market-based economies. Goods and service production are mutually interdependent. We believe, however, that services often play a more prominent role in this restructuring than this approach has generally allowed. Certain important aspects of service growth, recognized in the 'conventional' literature, need especially to be emphasized. These include the sometimes leading role taken by services in creating wider change, especially by providing key expertise, and the consequent tendency of service growth to create inherently uneven patterns of development.

We saw in Chapter 2 that in recent years the advanced economies have undergone significant changes, including (a) the internationalization of economic activities; (b) the reorganization of dominant firms; (c) the increasing integration of manufacturing and service production; (d) the growing use of micro-electronics technology; (e) the growing demand in industry for a highly skilled workforce, but with many routine jobs being displaced by technical change; (f) the increasing complexity and volatility of consumption; and (g) a changing role for state intervention. Earlier in this chapter we have seen these changes interpreted as a shift from a Fordist society based on large-scale mass production and consumption, supported by government demand management and welfare spending. Newly emerging forms of production, often described as post-Fordist, were identified where industry utilized new technology and a more 'flexible' workforce to respond more rapidly to market change and international competition, encouraged by a slimmed down and more entrepreneurial style of government.

We would in addition also emphasize the growing prominence of services and their significant and multi-faceted contributions to structural change, arising from; (a) the importance of the growing independence of goods and service production; (b) the value of service expertise in late twentieth-century capitalism; (c) the way technical change is creating new opportunities for the exploitation of service expertise and (d) the way service skills and expertise, embodied in the workforce, have a significant influence on locational patterns.

The interdependence of goods and services

Economic production has always combined material transformation and service functions but the distinction between goods and service production has become even more artificial in the modern era. Any material or service product is created by a complex sequence of material and service exchange. Those involved in this sequence, whether as suppliers or users, combine their in-house or self-service capacities with those of outside sources, including subcontractors and consultants. Supposedly 'manufacturing' corporations in many sectors increasingly focus on research, design, advertising and sales functions. They may simply assemble and market brand-named goods, exercising no more than quality and cost control over a manufacturing process undertaken by others. Sales strategies may be more important in ensuring profit than the cost of production itself, which is often trivial. In some markets for goods, such as food and clothing, the requirements of retailing corporations dominate manufacturing organization. The managerial and technical expertise of such service corporations nevertheless also depend on high-technology manufacturing innovation, especially in communication and control systems. All this demonstrates the misleading nature of the manufacturing-service distinction in any form of analysis, whether in the 'conventional' or 'alternative' tradition. Each type of function is fundamentally implicated with the other.

The importance of service expertise

The distinctive role of service functions lies primarily in the expertise they contribute to the manipulation of materials, information, capital or labour, in any production or consumption activity. Services should be defined, and their economic impacts measured, by the value of their specialist skills to their clients, within the wider division of labour. Such expertise need not necessarily be highly sophisticated, but must be beyond the capacities or inclinations of recipients. Many valuable services, such as repair work, gardening or decorating, simply offer experienced manual labour. Others enable materials, information or money value to be circulated through telecommunications networks of varying complexity. In the process, extra value is added, often from the advisory or interpretative expertise of the entrepreneur, the researcher, the technician, the doctor, the teacher or the sales person. The expertise content of services has become more important as the capacity for information exchange has grown. Services have thus not grown simply to circulate information but, as this has grown, to interpret the increasingly overwhelming volume of exchange. Even where services become more highly automated and impersonal, specialist expertise is still required, to design, establish and adapt such systems.

As interpreting the world becomes more complex, and manufacturing and much service production becomes more capital-intensive, the role of skilled service expertise thus grows. In our view, these conditions define the circumstances in which some service functions may initiate innovation and change. Walker's analysis gives significance to services only as 'indirect' functions in relation to material production, even though he acknowledges that they may be strategically very significant. We emphasize this latter characteristic and the autonomous qualities which some services may bring to bear on economic exchange, based in the value of human expertise. This certainly cannot be reduced to the material content of a computer disk, for example, whose value is negligible in relation to that of the information it may contain, and the uses to which this may be put. Similarly fast food restaurants may involve material production, but the qualities for which they are valued lie in the services upon which they rely, including their planning, financing and the design of the customer experience. Thus, while material production, and the technology it provides, certainly support many service activities, in many circumstances, the dependency operates equally in the opposite direction.

Innovation and technical change

Innovations involving information and communications technology are leading to new ways of exploiting service expertise, which in turn triggers growth and innovation across the economy. Barras (1986) notes that the balance of capital expenditure in US and UK services has shifted away from buildings to equipment, especially to technology based on micro-electronics. He believes this is reminiscent of the shift in investment towards factory production at the time of the industrial revolution. It reflects the start of what he calls a 'reverse product cycle' in service production, with the introduction of computer power initially leading to improvements in the efficiency of service production, then improvements in service quality and finally a range of new

Table 4.2 Barras's view of the reverse product cycle in services

Stage of cycle:	1. Improved efficiency	2. Improved quality	3. New services
Period:	1970s	1980s	1990s
Computer technology:	Mainframes	Online systems; minis and micros	Networks
Sector applications			
Insurance	Computerized policy records	Online policy quotations	Complete online service
Accountancy	Computer audit; internal time recording	Computerized management accounting	Fully automated audit and accounts
Local government	Corporate financial systems (e.g. payroll)	Departmental service delivery (e.g. housing allocation)	Public information services (e.g. videotex)

Source: Miles (1993).

services (Table 4.2). This view regards services as relatively passive recipients of innovation based on the introduction of manufacturing products. But rather than being technologically led, we see new technology as providing the possibilities for new ways of exploiting service expertise.

Information and communications technology can be used across a range of services and provide opportunities for back office and data processing automation, stock control and monitoring schemes in supermarket and physical distribution centres, more accessible public information systems, a range of financial services including automatic telling machines, smart cards and telephone banking services, and new forms of satellite broadcasting (Table 4.3). Such innovations are leading to:

1. the *industrialization of services*, i.e. the production of new services of predictable quality, capable of close quality monitoring e.g. fast food, franchising;
2. *organizational innovation* associated with technical change, e.g. supermarkets in retailing, self-service banking, information systems which facilitate more devolved systems of corporate organization, monitored and controlled centrally;
3. the linking of producer and consumer through technology which leads to new forms of *service trade*, e.g. telephone banking, sales and travel services.

Notwithstanding the highly automated outcomes associated with the application of information and communications technologies, in practice

Table 4.3 The characteristics of services and technical innovation

Service characteristic	Technical innovations
Service production	
Technology and plant Heavy investment in buildings.	Reduce costs of buildings by use of teleservices, toll-free phone numbers, etc.
Labour Some services highly professional (especially requiring interpersonal skills); others relatively unskilled often involving casual or part-time labour.	Reduce reliance on expensive and scarce skills by use of expert systems and related innovations; relocation of key operations to areas of low labour costs (using telecommunications to maintain coordination).
Organization of labour process Workforce often engaged in craftlike production with limited management control of details of work.	Use information technology (IT) to monitor workforce (e.g. tachometers and mobile communications for transport staff) aim for flatter organizational structures, with data from field and front-of-office workers directly entering databases and thence management information systems.
Features of production Production is often non-continuous and, economies of scale are limited.	Standardize production (e.g. fast-food chains), reorganize in more assembly-line like manner with more standard components and higher division of labour.
Organization of industry Some services state-run public services; others often small-scale with high preponderance, of family firms and self-employed.	Externalization and privatization of public services; combination of small firms using network technologies; IT-based service management systems.
Service product	
Nature of product Non-material, often information-intensive. Hard to store or transport. Process and product hard to distinguish.	Add material components (e.g. client cards membership cards). Use telematics for ordering, reservation, and, if possible, delivery. Maintain elements of familiar 'user-interfaces'.
Features of product Often customized to consumer requirements.	Use of electronic data interchange (EDI) for remote input of client details. In general, use of software by client or service provider to record client requirements and match to service product.

Table 4.3 The characteristics of services and technical innovation *continued*

Service characteristic	Technical innovations
Service consumption	
Delivery of product Production and consumption co-terminous in time and space; often client or supplier has to move to meet the other party.	Telematics; automated teller machines (ATMs) and equivalent information services.
Role of consumer Services are 'consumer-intensive' requiring inputs from consumer into design/production process.	Consumer use of standardized menus and new modes of delivering orders (EDI, fax, etc.).
Organization of consumption Often hard to separate production from consumption. Self-service in formal and informal economies commonplace.	Increased use of self-service, utilizing existing consumer (or intermediate producer) technology e.g. telephones; PCs and user-friendly software interfaces.
Service markets	
Organization of markets Some services delivered via public sector bureaucratic provision. Some costs are invisibly bundled with goods (e.g. retail sector).	New modes of charging ('pay per' society) new reservation systems, more volatility in pricing using features of EPOS (Electronic Point-of-Sale) and related systems.
Regulation Professional regulation common in some services.	Use of databases by regulatory institutions and service providers to supply and examine performance indicators and diagnostic evidence.
Marketing Difficult to demonstrate products in advance.	Guarantees; demonstration packages (e.g. 'demo' software shareware, trial periods of use).

Source: Miles (1993).

these innovations are critically dependent on the service expertise which devises and installs the equipment, and develops and manages the new forms of organization and service trade that go with it.

Service location

The leading significance of service expertise, embodied in the skills and capabilities of the workforce, has a significant influence on contemporary location patterns. In contrast to traditional expectations that service location tends to reflect the distribution of population and its purchasing power, modern service growth is inherently geographically uneven. In spite of the increasing ease of information exchange, the complexity and diversity of modern service expertise encourages agglomeration, at least of high-level functions. More routine functions may be more dispersed, although still controlled centrally. These trends have dominated the evolution of urban regions in recent years, and also now influence patterns of manufacturing location, as service expertise offers not just technical or material transforming expertise, but also organizational skills. The tendency for manufacturing to form new agglomerated concentrations, or 'new industrial spaces' (Scott, 1988a), probably arises less from the technical requirements of the manu-facturing processes themselves, than from their increasing dependence on specialist services (Wood, 1991a). The more complex, skill-based and market-oriented manufacturing becomes, in other words the more service intensive, the stronger the tendency towards regional agglo-meration.

Other aspects of service concentration affect regional patterns. Much exchange takes place between different service functions, and this has developed a growth dynamic of its own, as specialist service expertise is increasingly traded nationally and internationally. Thus, the business and financial services, with their surrounding media, cultural, travel, tourist or consultancy activities, are now powerful agents of economic, social and locational change. Consumption-related services, including retailing, tourism and recreation, and even public sector activities now significantly influence these changing regional patterns. Such services all depend on capital investment and complex material, capital and information exchanges. At the same time, the needs of geographically dispersed and variably mobile consumers have become more diverse. Thus, for example, retailing and health services focus their control functions into certain regions, as rising consumer demand supports new patterns of concentration of local service delivery.

Our 'service-informed' view of the economic consequences of modern restructuring thus does not separate service developments from wider structural changes in the operations of modern capitalism. It emphasizes that specific types of service function, offering particular expertise, participate in different ways in structural change. They sometimes lead and sometimes follow initiatives from other sources of economic and social transformation, at different times and places. The particular significance of service expertise needs to be evaluated in such cases, rather than being caricatured by old-established myths which give prominence either to manufacturing or to particular facets of service change. We shall return to consider the implications of this perspective for regional and local economic development in Section II, especially in Chapter 9.

Further discussion and study

Guided reading

Allen J, Massey D 1988 *The economy in question*, Sage/Open University, London: Chapter 3

Harvey D 1989 *The condition of post-modernity*. Edward Arnold, London: Chapter 9

Sayer A, Walker R 1992 *The new social economy: reworking the division of labour*. Blackwell, Oxford: Chapter 2

Walker R 1985 Is there a service economy? *Science and society* **40**: 42–83

Activity

1. Carry out a case study of a small number of manufacturing and service firms. Ask each firm about recent changes in their markets, the technology used in production, the externalization of functions they used to undertake themselves, the numbers and types of workers and their location.

   Are there any changes which are compatible with a shift from a 'Fordist' to a 'post-Fordist' form of business organization?

   What differences are there between manufacturing and service companies in their description of change? Does any shift away from Fordist forms of organization apply more to manufacturing or to services?

Consider

1. The transition from an earlier 'Fordist' pattern of production to a 'post-Fordist' regime is usually associated with changes in manufacturing. However, it also incorporates many other aspects of economic and social transformation. What are these, and how are private and public service activities implicated in them?
2. Compare the ideas of Gottmann and Walker on the urban and regional significance of service activities.

Discuss

1. As communication innovations appear to make the world smaller, does the location of economic activity become less important for its economic success?
2. The shift to a more service-dominated society since the 1960s has been associated with increased economic instability? Why?
3. The financial services (a) match the global supply of and demand for vital capital investment, but also (b) direct a growing volume of surplus capital into increasingly unstable speculative markets. Debate the changing balance between these roles and its effects during the 1980s.

4. To what extent is the urban and regional development of services dependent on manufactured products?
5. Compare and contrast various Marxist-based perspectives on service location and change.

SECTION II

Economic and social perspectives on service development

Chapter 5

Service work

Service work in the formal economy

Our service-informed analysis of modern industrial economies begins with service work. The most socially significant feature of service activities, especially over recent decades, is that they have created jobs (Gershuny and Miles, 1983; Gershuny, 1987; Graham *et al.*, 1989; see Chapter 2). Formally paid work has a central significance in modern society. For the majority of males it forms the basis of their social contribution, around which much of the rest of their lives revolves over a 50-year period. Increasingly, females also participate in the workforce for much of their lives. What people can buy and save is determined largely by the job they do, so that home and social life are shaped by the incomes associated with work. Jobs also confer status in society.

'Service work' has many connotations, its status reflecting the social context within which it is undertaken. In the formal economy, service workers include doctors and lawyers, teachers and politicians, management consultants and bank managers, airline pilots and check-in clerks, social workers and shop assistants, restaurant waitresses and bus drivers, street cleaners and chamber maids. Any such list demonstrates not just the variety of service work, but also the varied popular response to the idea of 'service'. To be a 'servant' is sometimes viewed as demeaning, perhaps associated with the servitude of household labour or exploitative casual work. The 'professional' services on the other hand, with their reassuring codes of ethical practice, have status, mainly because we depend on the expertise they offer. Nevertheless, attitudes may vary in different circumstances. Compared with the UK, manual service work carries less of a stigma in North America, especially if it leads to self-betterment, for example, by establishing successful small businesses, or by allowing students to work through university or college. On the other hand, doctors or lawyers who appear to be motivated by personal gain lose professional respect, even if envied for their commercial success.

Service work is rarely performed on a purely individual basis. It is dominated by the many agencies which organize the production of services; large and small commercial companies, government departments or public corporations, and professional bodies. The operations of these agencies, including some which extend across national boundaries, are examined more fully in other chapters (Chapter 6–8), but in general the motivations which drive them, whether concerned with profit, politics or professional status, also shape the nature of service work. In recent decades they have brought about radical changes in the nature of service work, and in the size and organization of the workforce. The readings in this chapter by Christopherson and Crang and Martin show how the following three interrelated trends have dominated these changes in both the private and public sectors.

The growing significance of service skills and high income services

The fastest employment growth has tended to be among high earners, reflecting increased demand for services offering specialized knowledge, frequently associated with new technologies (Collier, 1983; Urry, 1986) (Boxes 5.1 and 5.2). Although this demand is dominated by large private and public organizations, it is satisfied by both large and small private companies. Professional associations promote the interests of various groups of specialist service workers, for example in the medical specialisms, the law, accountancy or engineering, devising and controlling systems of accreditation. These associations both protect their members' interests and set standards, reassuring clients about the quality of service 'products' which often cannot otherwise be judged in advance. For many such services, including commercial activities such as financial services, there is also significant state regulation of the behaviour of workers, usually to protect users of the service (Rajan and Fryatt, 1988).

The standardization of much routine work, reducing reliance on human skills, and thus on many intermediate grades of service employee

This trend has depended on capital investment in technology, reducing and simplifying labour requirements for many established services such as retailing, office work and banking (Wood, 1989) (Boxes 5.1 and 5.2). New forms of routine service have also been created, for example in fast-food restaurants, or based on data processing, e.g. remote typing services or tele-cottages. Reduced reliance on workers' acquired skills has increased the significance in recruitment of their gender, age, ethnic or other personal characteristics, especially in customer-contact services, where established social stereotypes are exploited to create particular images of service quality (Santos, 1979).

Box 5.1: Assessing the impact of technological change on service employment

This chapter emphasizes how service activities have created significant numbers of jobs, especially during the last two decades. But it ends by questioning whether this will continue in the future. Technological change, and especially the increasing use of microelectronics-based computer and telecommunications technology, will have an important influence on both the quantity and the quality (see Box 5.2) of jobs created in services.

Assessing the impact of computer and telecommunications technologies on service employment is fraught with difficulty because technological changes do not have an independent impact; they are influenced by political, economic and social factors (Rajan, 1984, 1985, 1987a,b; Rajan and Cooke, 1987). Technological enthusiasts too readily assume that technological possibilities will come to fruition (Forester, 1987). Different methods for evaluating the impact of technologies also produce different results (Robertson *et al.*, 1982). The most comprehensive assessments of the impact of technologies place technical developments in the context of broader economic and social developments (Mandel, 1975; Gershuny and Miles, 1983).

To assess the impacts of technology on the quantity of employment in services it is useful to distinguish between differing types of technical changes. Microelectronics technologies may be associated with the following:

1. The introduction of a *new service 'product'*, e.g. a computer which facilitates the introduction of a new type of bank account or insurance policy (Rajan, 1987a, 1985). This creates new jobs associated with the development of new services, but jobs may also be lost if other activities are displaced (Thomas and Miles, 1986; Miles, 1988; Miles *et al.*, 1989).
2. New technologies may also be introduced as *process innovations* which lead to more efficient ways of conducting existing work, e.g. the use of the word-processor to do typing in the office. Though jobs may initially be lost as a consequence of the introduction of new process technology, if the costs of services are reduced, or new services made feasible, jobs can also be created.
3. Finally *managerial or administrative innovations* may introduce new ways of communicating in business or government, for example, through the use of video pictures transferred by telecommunications. Here, the direct impact of technologies on employment is less clear, but technology may be used to support new forms of business organization (see Chapter 6).

The distinction between these three categories of technical change is never clear-cut. But the scare stories associated with the negative consequences of computer-based technologies, common in the late 1970s and early 1980s, concentrate on only the capacity of technologies to automate existing work and thereby displace labour (Barron and Curnow, 1979). This must be balanced against services created by the technologies generating new jobs, perhaps in new industries (Gershuny and Miles, 1983; Barras, 1985; Miles, 1988). Such employment losses and gains are unlikely to occur in the same places, however. New industries tend to have differing locational requirements from established sectors. Innovations in communications also provide new spatial choices for corporate investment, by breaking down spatial constraints on locational decisions. Thus, technical changes are associated with considerable disruption of the economic landscape. Box 5.5 shows in two case studies the different employment and locational changes which can occur.

Box 5.2: The impact of technological change on service work and skills

Information technologies have differing impacts on the quality of work provided in services.

1. A pessimistic view, going back to early automation studies in the 1950s and 1960s, suggests that because automation can be used by management to control workers ever more closely, it may reduce the discretion required to do existing jobs, converting them to little more than closely monitored data entry tasks (Braverman, 1974). There are examples of such outcomes in the clerical factories associated with early versions of centralized computerized data processing (Crompton and Jones, 1984).
2. Alternatively, computerization may have a positive impact because it eliminates many routine tasks, and requires more skilled technical and professional workers to run, develop and manage the new technology (Forester, 1987).
3. It is also suggested that newer, more flexible, computer technology, because it distributes computer power to a range of functions (rather than tieing it up in a specialist computer department), may encourage a greater variety of outcomes including service innovation and job enlargement and enrichment (Noyelle, 1986; Rajan and Cooke, 1987). Workers may be moved from the performance of repetitive, boring assignments to work involving broader responsibilities, requiring a greater knowledge of the employing organization, its policies and its customers. At the same time previously separate tasks can be integrated (e.g. customer enquiries and data processing) by using the technology, allowing staff to develop new skills. These trends parallel those by which responsibilities are being pushed further down corporate hierarchies, making new demands on labour, and bringing about a new emphasis on the need for a better educated and better trained 'polyvalent' worker (Bertrand and Noyelle, 1988).
4. It is generally recognized that the widespread introduction of microelectronics technology will require the retraining of large numbers of workers to use it (National Economic Development Office, 1985).
5. It is important to keep in mind the segmented and low-grade character of much women's work when assessing the impact of technical change (Table 5.2, Fig. 5.1) (Phillips and Taylor, 1986; Walby, 1989; Pinch and Storey, 1992 a,b; Pinch, 1993). Microelectronics technology reduces demand for routine office administration, clerical and routine retailing tasks where female work is heavily concentrated. On the other hand, microelectronics appears to offer fewer new opportunities for women than men. Many jobs created in software development, database management and technical fields related to the technology are in occupations with relatively few women in them (Werneke, 1983). Women may lack both the technical background to gain the new jobs and access to such skills through the internal labour markets of companies. Also, when existing jobs are upgraded as a consequence of the introduction of technology, they are all too frequently redefined as male work (Jensen, 1989). It is thus likely women will bear much of the burden of adjustment to the introduction of information and communications technologies.

The growth of more 'flexible' forms of work

For standardized services, the employment offered is increasingly impermanent, insecure or part-time, and often predominantly for women, who have made an increasingly significant contribution to service employment (Table 5.1) (Curson, 1986; Hakim, 1987; Christopherson, 1989). Growth in contracting-out helps casualize the

Table 5.1 Women's participation in the labour force in OECD countries

| | Share of labour force | | Share in services | |
	1960	1985	1965	1985
Australia	21	38	70	81
Austria	34	39	na	68
Belgium	23	38	67	84
Canada	20	43	76	74
W. Germany	34	39	50	61
Finland	41	48	48	72
France	28	43	na	59
Italy	27	36	40	54
Japan	39	40	43	61
Norway	23	43	66	83
Sweden	33	47	68	83
Switzerland	34	37	61	73
UK	30	41	64	74
US	26	44	75	76

Source: UNCTAD (1989).

workforce (Allen and Henry, 1994). Contract work can also be used to undermine the wages and conditions of employment of established workforces. Under these circumstances, employers may easily adjust their labour requirements in relation to seasonal, weekly or daily demand fluctuations and, in the longer term, to new technological opportunities. For the higher-skill services, the principal manifestation of flexibility is the growth of self-employment, short-term contracts and consultancy services. This gives clients access to specialist skills as they are required, without having to cover all of the overhead costs of developing them.

Skilled managerial, technical and professional workers have not simply grown numerically, but have also become socially and politically more influential, even to the extent of being regarded as a 'service class' (Abercromby and Urry, 1983; Crompton and Jones, 1984; Urry, 1986; Savage *et al.*, 1988). Such social trends have supported a variety of other phenomena, including the rise of new social movements, urban and rural gentrification, greater environmental concern, the growth of cultural facilities and the arts in cities and the development of new patterns of recreation and lifestyle, all reflected by advertisers and the media. Pressures from the 'service class' for lower taxation, and their privatized forms of consumption in education, health care and pensions are also important factors behind government's attempts to control public spending and rationalize public services (Chapter 8).

In contrast, the organizers of routine services seek a pliant, non-unionized, dependent workforce which can be freely manipulated between alternative work arrangements. Service work has always been gender divided, with different functions regarded as especially appropriate for men or women (Leidner, 1991). Women have been consigned to less-skilled service occupations such as low-grade clerical

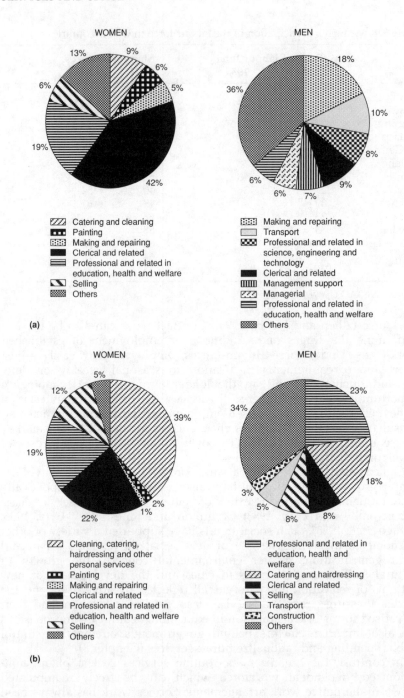

Figure 5.1 Great Britain (a) distribution of full-time waged workers by occupational grouping, 1986 (b) Distribution of part-time waged workers by occupational grouping, 1986 (Source: McDowell, 1989)

work, cleaning and catering or caring jobs in wealth and education (McDowell, 1991; see Fig. 5.1). In spite of pressures for equal opportunities, these and other stereotypes have been reinforced by the standardization of much service work (OECD, 1983).

Formal and informal service work

The web of social attitudes, and of power relations with clients and employers, which surrounds service work extends beyond the formal sector (Handy, 1985; Wood, 1989; Box 5.3). In fact, it has been estimated that this accounts for little more than half of all the service work undertaken in modern societies. Even in the commercial world, 'informal' social exchanges, within and between firms, are often seen as the key to business success (O'Farrell et al., 1993). The bartering of services or goods, and much technically illegal service provision, bypassing the taxation system, may well have become more common since the 1970s, in an era of growing unemployment and under-employment, mirroring long-established practice in developing economies (Clutterbuck, 1985; Redclift and Mingione, 1988; Portes, et al., 1989).

It is in domestic settings, however, that non-formal, unpaid work is of particular significance. Again, a high proportion of this is undertaken by women (Table 5.2). It provides a vital component of many consumer needs, thus influencing demand for formal service work. Such needs as feeding, cleaning, child-rearing, the maintenance of shelter, recreation or local travel are never likely to be satisfied solely through formal market transactions. In the past, the domestic division of labour seemed quite clear (McDowell, 1989). The classic mid-twentieth century family household consisted of a male wage-earner supported by a female 'housewife', supposedly dedicated to domestic service work. Male domestic responsibilities were conventionally confined to repair work and gardening. Consumer products, and many commercial services such as retailing and advertising were oriented towards the supportive role of women in this patriarchal arrangement, sustained by the earning power of male labour. This model did not apply to all households, especially among the poor where women also often had to take outside work. Its influence, nevertheless, gave general precedence to waged income, and undervalued the economic contribution of domestic, in relation to paid work, even though it was an essential complement to the latter.

Much has changed since the 1950s, and patterns of interaction between formal and domestic work are nowadays much more varied and complex (Pinch et al., 1989; McDowell, 1983, 1992, 1993; Pinch and Storey, 1992a,b,c). Many households do not fit the traditional family model at all. More single parents, older people and young singles place different demands on the public sector, informal and commercial services (Little et al., 1988). Many families contain two earners. This may be by choice, but often reflects the decline of secure male employment,

Box 5.3:   Different ways of looking at work

[Those] concerned with employment and labour market behaviour have rarely considered in any detail the domestic work of the household. Men and women are often plucked out of the complete context in which they do the various forms of work needed to get by.   Pahl (1984: 139)

Gershuny (1985) demonstrates the importance of 'self-service provision' in households. Services such as shelter, washing, entertainment, recreation and transport can be provided within households, using goods such as the TV, video, washing machine and car, or from the formal economy via the cinema, the laundrette or public transport. Shifts in service provision between the household and the formal economy, in response to social, legal and technical changes, have a significant influence on the growth of employment in personal services.

Gershuny and Pahl (1979) and Pahl (1984, 1988a,b) show how service work takes place in different spheres, including not only the formal economy, but also the household, a 'hidden' or 'underground' economy outside government statistics, and in the voluntary sector. It is necessary, they suggest, 'to enlarge our notion of work to embrace a much wider set of activities' (Pahl, 1984: 128). Pahl (1984) illustrates how the work undertaken by a woman ironing may be differently interpreted. She may be self-employed, or an outworker for a firm, or doing a task for herself, her husband or for a charity. Each type of work enmeshes the same task in different social relations; with different characteristics of production and a different balance of power between the woman and others. If, for example, the woman was ironing a garment for her husband she might be constrained to do so by conventional patriarchial values. Outworking for a firm in the formal economy or, still more, in the 'black' economy, might open the woman up to exploitation because of low wages and poor conditions of employment. On the other hand, she might be doing the ironing voluntarily for herself, or for a local charity. This choice of a woman in Pahl's example is not a coincidence. Women do much of the work in the household, and the informal sector more generally as well as increasingly in the formal economy. Their contribution to work is underestimated when only the formal economy is considered.

Individual and household work strategies are complex, involving different spheres of work. Again, using the example of a damaged roof, Pahl shows how differing sources of labour may be involved in its repair. It might be repaired either directly by the owners themselves, or indirectly through the landlord where a tenant lives in the property. Either might hire a contractor to do the work, thereby adding to income and employment in the formal economy. Or they might engage differing degrees of help from family, friends or neighbours. In these cases, though a payment in cash or kind could be made, the owner relies largely upon the informal economy to get the service done, with only the materials bought in the formal economy.

During the 1970s and early 1980s speculation grew that the unemployed might find alternative forms of rewarding work in the informal economy. It was even believed that informal work might be concentrated in deprived inner city areas (Robson, 1988), and that this might mitigate the impact of rising unemployment (Clutterbuck, 1985). Pahl shows, in his empirical work on the Isle of Sheppey, that unemployed individuals and households carry out informal work. But relative to those in formal employment, the unemployed have less skills, opportunities, equipment and capital to carry it out. In addition, by its very nature, informal work is unregulated, and provides employees with few rights and little protection. Inequalities in the formal economy are strongly reflected in informal work as well (Williams and Windebank, 1993). So, 'informal work reinforces rather than reduces the social and spatial inequalities produced by the formal economy' (Williams and Windebank, 1992: 19).

Table 5.2 GB household division of labour: by marital status, 1984 (%)

| | Married people* | | | | | | Never-married people† | | |
| | Actual allocation of tasks | | | Tasks should be allocated to | | | Tasks should be allocated to | | |
	Mainly man	Mainly woman	Shared equally	Mainly man	Mainly woman	Shared equally	Mainly man	Mainly woman	Shared equally
Household tasks (percentage allocation)‡									
Washing and ironing	1	88	9	–	77	21	–	68	30
Preparation of evening meal	5	77	16	1	61	35	1	49	49
Household cleaning	3	72	23	–	51	45	1	42	56
Household shopping	6	54	39	–	35	62	–	31	68
Evening dishes	18	37	41	12	21	64	13	15	71
Organization of household and money bills	32	38	28	23	15	58	19	16	63
Repairs of household equipment	83	6	8	79	2	17	74	–	24
Child rearing (percentage rearing)‡									
Looks after the children when they are at home sick	1	63	35	–	49	47	–	48	50
Teaches the children discipline	10	12	77	12	5	80	16	4	80

* 1120 married respondents, except for the questions on actual allocation of child-rearing tasks which were answered by 479 respondents.
† 283 never-married respondents. The table excludes results of the formerly married (widowed, divorced or separated) respondents.
‡ Don't knows and non-response to the question mean that some categories do not sum to 100 per cent.

Source: McDowell (1989).

as well as the increasing availability of part-time jobs disproportionately taken by women (Robinson and Wallace, 1984; Robinson, 1985; Dex, 1988; Bagguley, 1991; McGregor and Sproull, 1992). Such female participation in the workforce reflects their continued role in the household (Pinch and Storey, 1992b). Their domestic service responsibilities have continued, associated with child-rearing, caring for the elderly, cleaning, shopping and cooking (Table 5.2). This makes it difficult for women to work continuously full-time. Nevertheless, women are increasingly acquiring educational qualifications and taking up professional and managerial positions (Crompton and Sanderson, 1990). They have thus become dependent on new convenience-oriented household appliances and commodities to complete domestic tasks. There has even been a revival of domestic service work for the well-off double income family (Lowe and Gregson, 1989). Dependence has also grown on formal educational and welfare services to support child-rearing and household caring. In Europe especially, many of these services are provided by the state, and women are frequently employed to perform them. This means that women, because of their dual role in the formal and domestic economy, are in the forefront of the changes in the public sector discussed in Chapter 8. Such changes may introduce contradictory pressures, with women on the one hand being encouraged to work in the formal economy, while restructuring of the welfare state is predicated on women performing their traditional domestic caring role for sick relatives (Bondi and Peak, 1988). Clearly, then, changes in the formal and informal sector are intimately related (Pahl, 1984; Gershuny, 1985; Box 5.3).

So, how should we evaluate service work? It is evidently implicated in every aspect of modern life, and central to all processes of social change. If the value of service work depends on the benefits conveyed to others, the most important services should be those which cause the most socially significant changes in recipients. These may include mothers bringing up children, specialists saving patients' lives, local authorities housing homeless families, market consultants reviving the fortunes of companies, or brokers advising clients on profitable investment opportunities. The value of such disparate activities evidently cannot be compared directly, but much of the essence of modern society lies in these exchanges. As we have seen, the worth given to different types of service work reflects wider social attitudes. In general, for example, services which offer scarce intellectual expertise are socially valued, compared with activities based on manual skills. Yet in the modern world, we still depend on transportation or maintenance services, and are often more reluctant than ever to undertake more menial tasks ourselves. Old attitudes towards 'service' die hard.

The most powerful influences on the perceived value of service work in capitalist societies are commercial, derived from formal economic exchange. The changes in patterns of service work outlined above reflect these priorities, even in the public sector (see Table 4.1 and Chapter 8), with an emphasis on maximizing returns on capital investment, especially in new technology (Table 4.3). These criteria, of course, may reduce the costs and improve the quality of many private sector services, but are less likely to apply where non-commercial service qualities are valued. The worth of old town centres or rural landscapes, under threat from retail development, for example, or of health or education services, are unlikely to be reflected in such criteria. They also do not take account of externalities, such as increased unemployment or traffic generation, associated with private sector gain. We have also seen that commercially driven shifts in service work interact, in turn, with domestic and other informal service work. They thus drive wider patterns of social change, both through the labour market, for example in the increased demands being placed on women, and through their impacts on the balance of formal and informal work supporting consumption (Chapters 7 and 8).

The division of labour

This chapter has so far described service work in the broadest sense, whether in a commercial, public sector or domestic setting. These settings are interdependent, with the changing patterns of specialization within and between them underpinning much social change (Sayer and Walker, 1992). Over time, commercial pressures on services have made them more capital-intensive, both to cut unit costs and to deliver them to larger and geographically more widespread markets. Such pressures have primarily affected private sector services, such as retailing, banking and tourism, but are increasingly affecting public sector activities, including health provision, or local government, as these are opened up to competition. While economizing on routine labour inputs, these structural changes also place increased reliance on high-quality human expertise in specialist support services, and those functions within the organization that monitor competitive changes and make critical commercial judgements. Thus, those services which do not appear directly to involve high skills, still depend on them to support the capital investment which produces routine work, whether in the office, the retail store, the bank or the hospital (Chapter 4).

When viewed in aggregate, this changing structure of specialization, interdependence and growth is best understood as part of the evolution of the 'division of labour', described in Chapter 4 as a key element in 'alternative' approaches to the explanation of service growth. Patterns of specialization are found in all human societies. Under industrial capitalism, the principle was classically applied to manufacturing production to demonstrate the efficiency of factory compared with craft production. The joint application of complementary, specialized labour inputs enabled machinery to be used more efficiently than when workers had to perform many functions. A progressively more elaborate division of labour results from productivity pressures on labour resources. Although originally developed to explain the evolution of industrial specialization, such pressures have evidently expanded and diversified the service sector in recent years, and are likely to dominate the future of service employment.

The patterns of work specialism explained by the division of labour have a particular significance for service work, for at least three reasons:

1. As we have seen, lacking a directly tangible 'product', the contribution of service work must be validated through its relationships with other work activities, both formal and informal. The division of labour provides a framework which emphasizes these relationships.
2. Some have seen the growth of labour-intensive service work as a threat to the productivity of economies (Baumol, 1967). If the modern growth of specialized service occupations is part of an extending division of labour, this is not necessarily so. On the contrary, progressive work specialization is a means by which capitalism copes more efficiently with complexity.
3. The functional basis of the division of labour suggests that particular types of activity, produced apparently in the same manner, may have different social significance. Management expertise, for example, can be directed either to commercial, public service or charitable ends, each with distinct goals and criteria of success. The value of education and training depends on the type of recipient, whether it is 'produced' at school, in the home or at work. Banking expertise has a different significance in home and overseas markets, or for businesses and private individuals. The division of labour may serve the efficiency goals of providers, but work must also be tailored to specific social needs, reflected in client demand.

The internal structures of large commercial and public organizations most clearly demonstrate the participation of services in the increasingly complex twentieth century division of labour (see Chapter 6). Large organizations are dedicated to coordinating the work of many thousands of employees (sometimes spread across the world), with a diversity of specialist skills, towards the achievement of corporate goals. Specialist service work within the exploitation of this 'technical', or 'detail', division of labour has been of growing social and economic significance, as large organizations have gained domination over modern life. As well as segregating and coordinating skilled workers, we have seen that these processes also involve cutting costs by automating workers' tasks, in factories and also in service functions, such as in retailing, transport (larger lorries, automated warehouses) or clerical work (data manipulation, copying machines, word processing).

Organizations also specialize, however, into particular industrial sectors, or increasingly into areas of service expertise. This broader, 'social division of labour' explains some of the aggregate changes in patterns of service work, shown in Chapter 2. Employment in some commercial sectors is static or declining, especially where technical innovation has automated jobs, while that in others is growing, characterized by increasing demand. Specialist public and welfare services grew rapidly up to the late 1970s. Since then, in spite of attempts to reduce them, or to transfer some into the private sector, their employment role has more or less stabilized.

The growth of specialist business and financial service organizations has exploited the scarcity of expertise, in an increasingly complex and rapidly changing world (Chapter 6). It also marks an increased interdependence between the detail and social divisions of labour; between the specialist functions performed by firms' own labour forces and outside expertise, bought in when they cannot provide it for themselves. This has been encouraged during the 1980s by the rationalization of employment in many large organizations, as they focused on their 'core' expertise, and used outside advice for specialist technical, managerial, financial, marketing or legal problems.

The division of labour thus refers to the orchestration of all work functions in relation to socially defined needs, including those of capitalist enterprise and the overlapping role of state institutions, and communal and domestic activities. As Sayer and Walker comment, 'This extraordinary combination of separation and interdependence is so much part of contemporary life that we are thoroughly inured to it' (1992: 5–6). A rich array of arrangements is involved, linking organizations, workplaces ('from family to factory') and territories. All technological or management change results in changed methods of industrial organization, with resulting shifts in work practices and patterns of specialization. Indeed, with growing complexity, more expertise is being devoted to managing the division of labour itself, reflected in the growth of 'human resource management' functions.

The division of labour has profound social effects, since, as we have seen, occupational differences between people overlap strongly with gender, class and racial inequalities (Sayer and Walker, 1992). Work specialization is not socially neutral. It takes place within a system of power relations based on control, hierarchy, domination, persuasion or reciprocity. For some, positioned within the 'service class', the possession of scarce, high-quality service expertise, including know-how in financial markets, experience of managing businesses, legal and accounting skills, has become almost as powerful a route to success, as the control of capital. In fact, such expertise has become a major basis for capital accumulation (Thrift and Leyshon, 1992).

At lower levels of expertise, the requirements of service employment often depend on personal attributes. Much attention is paid by service managers to the need to attract and train workers who will represent service organizations effectively to clients. For good recruits, this may lead to effective career progression and relative job security. The process also leads to the common stereotyping of particular roles to female or male workers, or to the young or old, often to the disadvantage of women, older men, or ethnic and other minorities. Increasingly also, the performance of many customer-contact service jobs is subject to rigorous monitoring, so that job security is constantly under threat (Crompton and Jones, 1984; Box 5.2).

For many service workers, therefore, the division of labour is exploitative, with recipients, including employers and customers, gaining greater rewards than the provider. The growing inequalities between the best paid and most secure and the worst paid and least secure in society are not confined to the commercial sector. They permeate public sector functions, particularly as a result of privatization, subcontracting and casualization trends in local authority, health and educational employment (Chapter 8). Parallel shifts have affected domestic and consumer activities (Allen 1988b; Box 5.3).

Trends in the division of labour have also polarized the quality of services available to the better-off, compared to those supporting the poor. The former are largely employed in well-paid commercial or professional sectors and have access to private education, pensions and medicine. These are complemented by an effective range of informal business and institutional contacts, and by responsive public sector support, for example in relation to environmental quality and planning. The poor, often dependent on low-paid and insecure service jobs or welfare support, rely increasingly on public welfare services, which have generally been in decline, and on the informal sector and extended family links, which have also been eroded by social changes and housing problems (Box 5.3).

The location of service work

Service work is largely an urban phenomenon, despite the capacity of telecommunications-based technology to support remote working (Miles, 1988). Service growth, alongside both the decline and the decentralization of manufacturing to less-urbanized locations, means that service employment dominates large metropolitan centres (Frost and Spence, 1993; Box 5.4), and even many smaller cities and towns (Chapter 2; Stanback *et al* 1991).

The two **readings** in this chapter (pages 104–18) give more detailed insights into the location of service work at different spatial scales: the national and the local. They show the different ways in which services contribute to disparities both within and between local economies. Christopherson's paper examines aggregate service employment trends in the United States, the largest and possibly most fluid high-income labour market in the world. She draws out similarities and contrasts between the geographical effects of modern work practices in the manufacturing and service sectors.

1. In both cases, large corporations increasingly dominate patterns of employment, although this is a more recent development in services (Gudgin, Crum and Bailey, 1979). Higher level jobs, in management, and research and development, have become increasingly concentrated into a few, usually metropolitan-based control centres. High skill-based business services, where there are also many small firms, are also located mainly in large cities, adapting to different patterns of manufacturing and service demand (Chapter 6).

2. The contrasts between the sectors emerge in the organization of routine work. It has been widely observed that modern, 'flexible' patterns of manufacturing work tend to *differentiate between locations*, exploiting their particular qualities, whether of cheap or skilled labour, the activities of subcontracting firms, the availability of greenfield sites, access to markets, or inducements from public agencies. In contrast, the flexible work practices developed by large service firms, for example, in retailing or private healthcare, mean that *many sites are broadly similar* in their structure of employment, since the purpose is to deliver the same service to as large a market as possible (Chapter 7).

Christopherson argues that this widespread standardization of routine service work is as important a consequence for employment as is the spatial division of labour in manufacturing. Its main implication has been the geographically widespread adoption of flexible working methods, enabling large service organizations, including federal government and other public agencies, to minimize costs by exploiting part-time, temporary and self-employed workers (see also reading, Chapter 8, pp. 197–204). This 'buffer' workforce can be adjusted in relation to temporal and geographical variations in demand. Much of the growth in this workforce is of women, who have generally been employed less in the US in the past than in Europe. For some, this is undoubtedly voluntary, but high male unemployment in many rural and old industrial and mining areas, and more widely among the young and ethnic minorities, means that women's work is becoming increasingly necessary to support household incomes. Christopherson argues that low levels of unionization and political power among this widely scattered workforce means that it has little choice but to accept the new forms of work that are being made available.

As formal service work has grown, so has the need to explore the social consequences of the relationships between trends in different service industries. The **reading** from Crang and Martin (page 111) illustrates the importance of these relationships, demonstrated especially within a particular geographical frame; the 'local labour market'. They chose a well-known case of service-based local prosperity, in a medium-sized town which became a high-profile example of economic 'success' during the 1980s in Britain. The 'Cambridge phenomenon' was widely promoted as a successful high-technology, tourist and university centre, following a report by consultants Segal Quince and Partners (SQP, 1985). The first part of the paper, not reproduced here, emphasizes that the high technology base of the city was not a guarantee of long-term success, nor could it easily be replicated elsewhere.

The sections presented here emphasize the 'other side' of the local employment impact of Cambridge's success in the 1980s, the routine consumer services referred to by Christopherson. They draw attention to the ways in which high-tech developments depend on low-paid cleaners and other support staff. They also argue that growing local consumer service employment, expanding as a result of growth in local

---

Box 5.4: Sassen's polarized 'global city'

In order to understand . . . major cities today, we need to examine . . . the world economy. . . . [C]hanges cannot be adequately explained merely in terms of the shift from manufacturing to services. (Sassen, 1991: 323)

Sassen, like Gottmann (Box 3.1), recognizes that services have replaced manufacturing as the economic heart of large cities. She shows that in New York, London and Tokyo during the 1970s and 1980s, the contribution of manufacturing to total employment declined by 5–6 percentage points. In 1985 manufacturing only employed 15–22 per cent of all employees, while service industries, which had expanded rapidly during the previous two decades, employed 59–78 per cent (Sassen, 1991). Unlike Gottmann's early work, Sassen argues that processes of global economic integration, and the increasing participation of service agglomerations in large cities in international trade, lie behind these changes (see Chapter 9).

Sassen thus links established literature on large cities (Hall, 1984, 1989) with that which sees them as part of a global economic system (Friedmann and Wolff, 1982; Friedmann, 1986). Her point of departure is that the increasing integration of the world economy has produced a rapid expansion of international financial markets, a growth in trade in professional and business services, and the increasing international operations of large manufacturing and service firms (see Chapter 9, especially Box 9.1). A small number of industrialized countries dominate this global economy, and within these countries a small number of large firms dominate trade flows. As a consequence, a few major cities, which act as bases for these large organizations and their support services, emerge as key hubs in the international economy. These 'global cities', such as Tokyo, New York and London, serve as internationally integrated centres of finance capital, and provide a location for business command functions controlling a network of sites across the globe (Sassen, 1991).

When manufacturing industry provided the economic base of large cities, it supported a broad middle class, including managers and administrators, as well as skilled blue collar workers. In contrast, the rise of the new service complex has contributed to growing economic and social polarization (Sassen-Koob, 1984). The new service industries in large cities have a polarized occupational and income structure; combining a high income and status professional and managerial 'service class', and large numbers of routine clerical employees, with lower than average incomes, and frequently working part time. The growth of high-income jobs and the consumption behaviour of such employees, also produces related growth of low-wage employment in restaurants, hotels, cleaning and laundry establishments. Employment in these areas frequently draws on a casualized 'underclass' prepared to work in an informal sector where they have few employment rights and are largely unprotected by employment legislation.

The urban structure of the city reflects this social polarization, and is marked by high-income gentrification and related conspicuous consumption by employees in the affluent 'service class', cheek by jowl with poverty, urban deprivation and run-down areas. Such contrasts are reinforced by the fact that public services, which traditionally supported the less well-off in the city, are being undermined by the privatized consumption behaviour of the better-off, and associated pressure for reductions in local taxation (see Chapter 8).

---

incomes, discriminates against some groups in the community, with particular class images, ages, aspirations and even from some areas of the city. The growth of each of the four pillars of the local service economy, business and financial services, tourism, consumer services and education, creates much subcontracted, servile work. The ambiguous position of women is also emphasized. In general, therefore, the extracts demonstrate how the service economy is polarized at the local level, even in a supposedly successful local economy (see Warf, 1989; 1990, and Box 5.4 for similar arguments with regard to large cities).

Such studies can trace the broad impacts of the growth and changing organization of formal service work. But they do not directly examine the detailed social consequences of these changes. For example, they only speculate whether the shift to flexible work practices is voluntary or involuntary, and what its effects may be on household incomes and wider social practices. Such questions require much more detailed, sociological inquiry. Pahl's (1984) work sets out some of the issues which operate at this individual or household scale (see Box 5.3). He draws on the pioneering work of Gershuny, arguing that various interdependent and mutually supporting 'spheres of service provision' are each important in providing for formal economic and domestic needs. His illustration of a woman ironing for a variety of different purposes shows how social context determines the significance of service work. Also, the achievement of any particular task, such as mending a roof, may involve various alternative social and economic arrangements. These examples demonstrate the various relationships of exploitation or dependence upon which all service work is based. Households possess some scope for informal adaptation in the provision of basic consumer needs, for example if the availability of full-time male employment declines, and dependence grows on insecure female employment. But these adaptations may place severe pressures on household relationships; secure sources of formal income are still fundamentally important. The character of the informal economy, including its spatial distribution (Williams and Windebank, 1993), is likely to be closely related to the formal, and informal work is not a substitute for formal employment.

The future of service work

A significant characteristic of much service work is that traditionally it has been labour-intensive (see Chapter 2). By the standards of extractive and manufacturing industries, service functions even in the formal economy have proved resistant to automation. Growing demand, especially in recent years, has thus been more readily transformed into employment than for materials-processing (Chapter 2). Low labour productivity, however, either tends to make services expensive for the consumer, or means that part of the workforce is poorly paid. With higher productivity, automated methods of provision can be cheaper and more profitable. For example, prices in the corner shop, or even the high street, cannot compete with those at the out-of-town supermarket. Fast-food outlets displace traditional popular restaurants. Cash points are taking over much of the routine work of bank clerks. Even in the business services, training programmes, for example, are increasingly becoming computer- or video-based. These trends are driven by national and even international competition between major companies, so that more expensive forms of local provision are no longer protected. Many service jobs therefore appear susceptible to the same processes of automation and displacement that have affected manufacturing employment in recent decades (Box 5.1).

Such competitive and productivity pressures challenge any assumption that service employment will continue to grow rapidly. Lester Thurow (1989), the American economist and columnist, has argued that, for this and other reasons, services 'are not the wave of the future', but of the recent past in high-income societies. For him the rapid employment expansion of the 1970s and 1980s was a passing phase during which growth in demand for consumer, producer and public sector services temporarily outran the automating capacity of new service technology. In the USA a high proportion of this growth followed 'one-off' changes, such as the shift to more female employment, the move to 24 hour, 7 day a week shopping, the commercial real estate boom and financial deregulation. The effects of these changes will not be sustained into the 1990s, when cost and competition pressures, harnessing especially the potential of communications and data processing technology, will cause labour requirements to fall in many services.

The projection of past trends should always be treated with caution. While service employment growth has been occurring throughout most of the twentieth century, the unusually rapid growth of the 1980s is unlikely to be sustained for long. Nevertheless, the growth of service work, as we have seen, is supported by a complex series of social and economic relationships. No single source of change, such as the automation of established services, can be the only influence on levels of employment (Box 5.1). Manufacturing employment is even less likely to grow rapidly, except in a few sectors developing new products. If formal service work declines, therefore, more people may be unemployed, but its overall contribution to the supply of work in the formal sector may still increase. The influence of various forms of informal service work may become correspondingly greater. Trends within the formal service sector, and between it and the informal sector will still dominate work patterns in the future.

The future scale of service growth, and the character of service work depend on the balance to be struck between a variety of trends.

Pressures towards service growth

1. Automation, while generally displacing less skilled workers, increases demand for the more skilled technical, organizational and financial management services, which are more difficult to automate. Such 'business services', whether in specialist service firms, or employed by manufacturing and other service organizations for their own use, will continue growing (Box 5.2).

2.   Automation also generates new services. Completely new types of business have developed on the basis, for example, of travel booking systems, mobile telephones, telephone sales networks and television and video programme production. Barras's (1985) 'reverse product cycle' theory identifies this as a general tendency, especially arising from the spread of communications and information technologies. As these become more routine and widely accepted, a plethora of new and specialist services are generated (Table 5.2).

3.   In the 1980s, the growing productivity of capital-intensive manufacturing and service activities *depended* to some extent on displacing some of the jobs they might have supported into separate, more labour-intensive services. This was one facet of the trend towards 'flexible' employment relations, allowing both private and public agencies to avoid employing workers directly, engaging them instead from subcontractors or consultants, on cheaper annual cost terms (Coffey and Bailly, 1992). This trend has affected private firms, including the health care companies Christopherson describes for the US. But the same patterns can also be discerned in local and national government agencies there, and in the UK, Australia, New Zealand and Sweden, spurred by privatization and the introduction of competition (Osborne and Gaebler, 1992). Although such changes are 'one-off' for individual organizations, they have yet to be adopted by many potential service clients. Service work in these activities is therefore likely to continue growing, even if at the expense of similar jobs in client organizations.

4.   Other types of 'flexible' work will also continue to increase, including more part-time, temporary and self-employed service work. In the US, emphasis has been placed on the polarization of opportunities accompanying the growth of service employment during the 1970s and 1980s. Many low-quality jobs were created, including some for males who would in the past have been employed in semi-skilled manufacturing. Such forms of service work may in future grow more quickly in Europe, where more publicly controlled labour markets have delayed the widespread introduction of flexible work practices, compared with the US. Different patterns of wage flexibility and social security systems will, however, influence the detailed patterns of response with women, for example more often engaged in flexible work arrangements than in the US (Pinch and Storey, 1992c). One implication from Crang and Martin's analysis (in the reading) of a particular labour market is that the expansion of business services and high technology activities stimulated demand for other forms of work, even if it has an inherently polarizing effect.

5.   Some services markets are dominated by state regulation, which may change in the future. Deregulation drove much of the growth in personal financial services during the 1980s in the US and UK economies. Similar growth potential still exists, for example, in

some sectors of air travel, telecommunications and satellite programme transmission in Europe. The negotiations for the single European market have expended much effort to reduce regulatory barriers to service trade and investment, and thus encourage growth. Conversely, some forms of regulation generate demands for new services. The best examples of this are probably in environmental and occupational health legislation, which are spawning a new range of specialist consultancies to advise on the practical and legal aspects of their implementation.

6. The range of specialist services demanded by both consumers and businesses should continue to expand. The corner shop has a role as a local 'open all hours' convenience store. The high street may specialize in quality goods and services, including specialist restaurants to compete with fast food outlets. Banks will need to offer more attention to specific customer needs. The specialized requirements which underlie the growth of business services will also continue to diversify (see Chapters 2 and 6).

Pressures towards service employment decline

1. Productivity pressures will undoubtedly mean that the numbers of many types of service jobs will stagnate or decline. By the late 1980s, the financial services boom in the UK and US economies had gone into reverse, with banks, insurance companies and building societies drastically rationalizing their networks, and cutting employment requirements. An important aspect of change in the financial sector, during the 1970s and 1980s, was its involvement in the global market. As other services increasingly participated in international trade, this not only opened up new business opportunities based around service specialization, but also reduced the scope for less efficient producers to shelter behind local and even national barriers to trade (Chapter 9; Petit, 1986).

    The spread of supermarket retailing, and the out-of-town shopping mall in Europe, as in the US, will reduce the total employment in transport and distribution. Whatever the general levels of economic prosperity, such pressures will ensure future service employment decline in such established activities, where the prospects for new service 'products' are likely to be limited. The security of much traditional, labour-intensive clerically based service work will also decline. Its growth in the 1980s, was for female and part-time workers (Pinch and Storey, 1992c). New opportunities will arise, for example in telephone-based customer services, but these will be in different parts of the country, and will involve the introduction of more flexible working methods (Box 5.5).

2. The constraints on the growth of public sector employment are likely to continue for the foreseeable future. Fierce international competition in tradable goods and services will also act to hold back public service growth which would add to business costs

Box 5.5: Case studies of the geographical impact of technical change

Using information from Richardson (1993), two case studies show how intelligent telecommunications networks, combining advanced telecommunications and computer facilities, may provide new telephone-based services, leading to a geographical restructuring of employment. Telephony mitigates the need for face-to-face contact between customer and service provider. The integration of telecommunications and powerful computer networks also provides on-line information which is used to provide new services, or target groups of customers more effectively for telephone sales. Richardson shows how the result may be decentralization of especially routine service employment from large metropolitan areas. In the first case study, telephone-based services are exported from a region chosen for its lower production costs. In the second case study, a company uses intelligent telecommunications services to decentralize employment, reduce costs and improve the quality of labour.

First Direct
Competition in British personal financial markets increased during the 1980s as building societies and banks entered each others' traditional markets. The Midland Bank was a high-cost producer with extensive debts. In 1989 it sought to reduce costs and gain access to more profitable high-income customers by launching First Direct, a new form of 24 hour telephone banking service. This provided customers with traditional banking services including cheque book accounts, credit cards, savings accounts, share deals and loans over the telephone.

First Direct was established at a new site on a business park in Leeds in northern England. This allowed the company to establish a new work culture which was less hierarchical and bureaucratic than the bank's other businesses. Women were selected as customer liaison staff as on the basis of behavioural skills, personality and telephone manner, rather than experience in banking. They were supported by a smaller group of male specialists who dealt with more complex business. The organization was sales-oriented with staff encouraged to 'cross sell' the bank's various products to customers. Staff performance was monitored by computer and an element of performance-related pay was introduced.

By 1993 First Direct employed 1450 people, and had 400 000 customers, many attracted from other banks. Customers were largely more affluent people in the South East of England. These were served from Leeds in northern England, which offered lower office and labour costs, without the need to expand a branch network on high streets in the South East.

British Airways
British Airways (BA) offers a range of booking services to travel agents and members of the public in the UK, including enquiries, sales and direct access to its own databases and computers. The vast majority of customers are in the South East of England and, until the late 1980s, they were served by telesales staff based at Heathrow, near London. As a result BA was burdened with a high-cost, high-turnover labour force, reflecting the wide range of alternative job opportunities close to the capital.

In the late 1980s BA reorganized its telesales, establishing regional centres at Manchester, Belfast, Glasgow and Newcastle. Employment at Heathrow was reduced from 900 to 250. The new telesales operation functions virtually as a single office. BA provides local call telephone numbers to its customers, which are routed automatically via private telephone lines to one of the regional centres, and then to another if the first is busy. Different telephone numbers are provided to different groups of customers (e.g. previous customers or top travel agents), and these calls are passed to separate parts of the regional network specializing in dealing with each group. The regional centres are monitored by top (cont.)

Box 5.5: Case studies of the geographical impact of technical change
(continued)

management based at Heathrow. Telesales groups dealing with key corporate customers are also based there. UK telesales has also been integrated into BA's world network, and, for example, 15 per cent of US calls are answered by UK telesales.

The new regional centres offer BA a reliable, well-qualified, stable workforce. This reduces training costs. Further savings in wage costs have been achieved by introducing regional pay bargaining, to take advantage of wage differentials between the South East and other parts of the country. Performance-related pay has also been introduced in the regional centres.

---

through taxation. While the continued growth of female participation in the labour force will add to demand for a range of services, many of which have traditionally been supplied by the state, these will be increasingly shifted to the private sector, or will incur higher levels of direct fee payments, so that cost will act as a constraint on demand.

3. More speculatively, the polarization of the employment structure encouraged by the growth of services may, at least partially, undermine the elements of the mass market upon which the growth of many services and also goods have depended since 1945. It is also possible that the impact of this change may be stronger in some areas than others.

4. Many households will thus be pushed towards greater dependence on semiformal and informal means of support, and the state welfare sector, to supplement low incomes from the formal sector. As Pahl has emphasized while *individuals* may earn only modest formal incomes, *households* may supplement these by intensified informal arrangements, to cope with reduced security. Trends in the nature and significance of informal work, will therefore take on a wider social significance (Box 5.3). The poorest, both relatively young and old, will be those who are unemployed, and have no other access to income through other household members.

5. In many circumstances, the weakening of traditional nuclear and extended family structures has also reduced support from informal service networks. Thus, the state or voluntary sectors have expanded to bridge the gap. Even though economic and political pressures may bear down on state service provision, wider social trends have therefore placed increasing demands on such provision to accommodate higher levels of need (Chapter 8).

Location trends

In sum, demand and productivity pressures on service work will offer more polarized opportunities in all areas, with success depending heavily on organizations commanding education and training skills, and heightened awareness of trends in potential client requirements. Such growth will be concentrated into some areas, where service skills interact and develop, and from which they can be widely traded. These have traditionally been in major cities, whose local labour markets have themselves become increasingly polarized as manufacturing has declined. Cities have become the centres of service-based economic restructuring on both sides of the Atlantic (Box 5.4). Many of their more labour-intensive functions, however, are in relative decline. More routine functions will continue to decentralize to outer metropolitan areas and smaller cities, to avoid the high costs of a city centre location, and this will often be accompanied by a growing flexibility in conditions of employment, and a net reduction in the numbers of full-time equivalent workers, as automated methods extend their influence (Box 5.5). The largest cities are likely increasingly to become the focus for employment in the higher echelons of large firms, in financial and business services and in the consumer functions attracted by their incomes and traditional roles as regional centres.

Further discussion and study

Guided reading

McDowell L 1991 Life without Father and Ford: the new gender order of post-Fordism. *Transaction of the IBG* **16**: 400–19
Pahl R E 1984 *Divisions of labour*. Blackwell, Oxford
Sayer A, **Walker R** 1992 *The new social economy: reworking the division of labour*. Blackwell, Oxford

Activity

1. Table 5.2 shows both the division of labour within British households in 1984 and attitudes to it.

    (a) Identify those tasks that are predominantly undertaken by males, females and shared.
    (b) Do the results reflect:
        (i)   the physical abilities of men and women;
        (ii) time availability, because of formal work obligations;
        (iii) social attitudes to what is appropriate for men and women to do?
    (c) Would household work be differently shared if the attitudinal responses in Table 5.2(b) were implemented? What conclusions do you draw about British adaptability to change in the nature and availability of work?

2. From the reading taken from Christopherson's paper (pages 104–11), identify various types of 'flexible' service work in the USA, noting their scale and composition. Why does she believe that service developments are reducing local differences between *job opportunities* and *consumption patterns*?

Consider

1. What are the 'back regions' of Cambridge, and who occupies them? (pages 111–18)
2. In Box 5.3 Pahl shows how the significance of service work depends on the social context in which it is performed, using the example of why a women might be ironing; as paid work, for herself, for her family, for a friend or for charity. Whether the work is being done willingly or under duress; to earn vital income, or to satisfy social expectations, are also relevant. Analyse some other tasks in the same terms, for example:

    (a) preparing household financial accounts;
    (b) repairing a car;
    (c) serving in a shop;
    (d) cooking;
    (e) digging a garden;
    (f) looking after children.

Discuss

1. Classical economists saw the *division of labour* as one of the chief processes by which industrial productivity was increased. Workers were given more specialized tasks, which were dependent on investment in more technically sophisticated machinery. It thus allowed producers to increase output, reduce unit labour costs and achieve greater economies of scale. Labour also became more dependent on capital for employment. How far does this drive towards progressively greater worker specialization and control explain the growth of various services?
2. What factors affect whether you might occupy a high-income or low-income service job? How far is the choice under your, and any other individual's, control?
3. Why have increasing numbers of women entered the labour market? Why have service employers especially preferred women employees? Describe the links between the gender division of labour in formal and informal work and any changes since the 1950s.
4. Assess how the character and location of service work will be affected by the increasing use of microelectronics technology in the workplace.

Readings

Flexibility in the US service economy and the emerging spatial division of labour

S CHRISTOPHERSON, 1989 (*Transactions of the Institute of British Geographers* 14: 131–43)

Rationalization and specialization: complementary patterns in services

... A wide range of evidence demonstrates that manufacturing is being carried out in smaller enterprises producing a wider variety of outputs. US service sector firms, however, are characterized by quite different trends in production organization. One tendency is that of systemic rationalization and a drive to achieve scale economies. The other tendency is toward specialization of service products. Even more confusing, these two patterns may exist within the same firm.

In contrast with trends in manufacturing, employment in the fastest growing US service industries is concentrated in very large firms which have expanded by increasing the number of establishments they control. The general direction toward production in smaller units (that has been noted as occurring across industrialized countries) is more substantially attributable to the increasing number of small units in service firms than the vertical disintegration of manufacturing (Loveman and Sengenberger, 1988). ...

The fastest growing industries in the US, business services, retail and health are clearly dominated by large firms. Over 50 per cent of retail employment in 1984 was in firms with over 100 employees. In consumer, health and business services 66 per cent of employment was in firms of this size and 46 per cent in firms with more than 500 employees (Small Business Administration, 1987, p. 279).

This pattern of concentration in large multi-establishment firms is most evident in the retail sector where the number of establishments in firms with the highest volume of sales has significantly increased (US Department of Commerce, Census of Retail Trade 1972, 1977, 1982). Although retail is still dominated by small establishments, the proportion of establishments in the lowest total firm sales category declined from 92 per cent of total establishments in 1972 to 85 per cent in 1982. This reorganization in large, higher sales volume multi-establishment firms is even more apparent in employment figures. Between 1984 and 1986 employment in retail firms with over 500 employees increased 16 per cent compared to an increase of only 2 per cent in firms with less than 100 employees (SBA, 1987).

In the health care industry, there is only indirect evidence but the same apparent pattern. For example, 28 'for-profit' hospital chains managed, owned, or leased 809 hospitals in the US in 1983 and 958 hospitals in 1984. Their profits increased 30 per cent during that year according to a report in *Modern Healthcare*, a trade journal. In both these industries, the trend is toward achieving scale economies by producing and distributing services through multiple market-proximate units.

In addition to multiplying units through which to produce and distribute services, large US service firms are taking a series of steps to standardize and rationalize production. As in manufacturing, there has been significant capital investment (primarily in computer technology) aimed at reorganizing the work process (Stanback, 1987). Secondly, the workforce has been redeployed between routine (low value-added) and non-routine (high value-added) work. For example, routine nursing care has been shifted out of hospitals to different types of routine nursing care facilities. As a consequence, the non-hospital health care sector grew 70 per cent in the 1970s. This new pattern of health care provision is exemplified by Beverly Enterprises, a nursing home chain which employs 116 000 people (more than Chrysler Corporation at 115 000) and deploys their labour in 1200 different locations.

The emerging spatial form of the service firm is quite different from that which we associate with manufacturing enterprises. First, these firms are characterized by the physical separation of labour among a set of worksites with similar functions (the retail outlet, the nursing home, the rental office). Second, even within the corporate headquarters, distinct worksites are emerging. In the health care industry, activities which once took place in the same building, such as out-patient care or laboratory testing, are being physically separated to allow separate billing. In addition, as financial management and marketing have become more important to the modern hospital, administrative staff has increased relative to health care workers. This work, too, is more frequently carried out under separate administration and frequently in a location apart from the hospital itself (Schoen, 1987).

Although employment in consumer and distributive services is concentrated in large firms, there is a parallel trend toward small firm growth in producer services or what might be called services to services. Case studies of vertical disintegration in service industries suggest that much of this growth is an aspect of the service rationalization process (Noyelle, 1986; Baran, 1986). For example, large firms more frequently subcontract activities which they formerly carried out (i.e. food services; laboratory tests; building maintenance). Whereas, in the large-firm, small establishment model, there is little specialization across units, subcontractors provide specialized services. These building maintenance and catering firms, too, are currently growing in size as a consequence of mergers, occasionally financed by the firms to which they subcontract their services.

Although they originate in two different processes – the vertical disintegration of production and the proliferation of small establishments within large firms – the down-sizing of the workplace and physical separation of production activities are among the most prominent aspects of contemporary service employment in the United States. Fifty-five per cent of all workers work in an establishment with 100 employees or less. What is important to recognize is that the large firm does not disappear in either of these processes. It merely assumes a different role, emphasizing finance and distribution, and distancing itself directly or indirectly from the actual production and provision of the service. The drive to achieve scale economies in service provision suggests that an expanded use of flexible labour in these industries is associated primarily with rationalization and cost reduction. It is only marginally a consequence of the need to produce multiple differentiated outputs. . . .

The flexible workforce in the United States
From 1973 to 1979, 12.5 million jobs were added to the US economy. Another 14.5 million jobs were added in the 1980s. Nearly a quarter of these jobs were *part-time* and approximately 66 per cent were filled by women. One out of every six US jobs or about 19 million total is a part-time job. This yearly average figure understates the dimensions of the part-time work experience, however, for a much larger proportion of the workforce is employed part-time at some point during the year. In 1985, for example, the number of people who worked part-time for a portion of the year was double that of the annual average number of part-time workers (Tilly, 1988). Eighty-nine per cent of the part-time workforce is employed in the service industries. Within service industries, the most important employer of part-time workers is the wholesale and retail trade where 30 per cent of the workforce is composed of part-time workers.

Part-time jobs are not only a sizeable portion of total employment, they are growing faster than full-time jobs. Of the 10 million jobs created since 1980, one quarter have been part-time. Part-time work is divided among two groups of workers, those who work part-time by choice and those who work part-time because full-time work is not available to them. Thirteen and a half million workers work part-time voluntarily while a growing portion of the part-time workforce, 5.6 million, are involuntary part-time workers.

Since the 1950s, part-time work in the US shows two trends: a secular increase over time and a tendency to fluctuate more with respect to the business cycle. Part-time work has increased very slowly as a portion of the total work force (from 16 per cent in the 1950s). This long term secular trend has been substantiated by studies which separate cyclical and secular trends (Ichnowski and Preston, 1985; Ehrenberg, Rosenberg and Li, 1986). The increased cyclical sensitivity of the part-time workforce is attributable to the larger portion of this workforce which is involuntary. In the mid-1970s the rate of growth in voluntary part-time employment began to slow and that of involuntary part-time employment to increase. Of the 2.9 per cent increase in part-time employment between 1969 and 1987, 2.4 per cent is attributable to growth in involuntary part-time work (Tilly, 1988).

In the 1940s, men part-time workers outnumbered women because of the concentration of part-time work in primary sector industries (Leon and Bednarzik, 1978; Morse, 1969). (Twenty-four per cent of the workforce in agriculture currently works part-time but agriculture is much less significant with respect to the sectoral distribution of employment.) The contemporary part-time workforce in the US is disproportionately composed of younger and older workers and of women in comparison with the workforce as a whole. Fifty-four per cent of part-time workers are wives or children in married couple families.

Although part-time jobs are the most numerous of flexible jobs other forms of flexible employment are expanding more rapidly, particularly, *temporary work*. The use of temporary workers is especially prevalent in those situations, such as general clerical work and data entry, where the pattern of demand is not predictable and in work where generalized rather than firm-specific skills are required. Temporary work contracts can take a number of forms including:

1. a short term job;
2. a long term job with no employment security, lower pay or no benefits;
3. a structured internal temporary worker pool (most common in large public institutions, such as universities and hospitals).

In most cases, recruitment and hiring of temporary workers is carried out independently of the hiring of permanent personnel. Temporary workers differ from part-time workers in two important respects. Part-time workers more frequently work variable hours (rather than generally full-time as do temporary workers) increasingly geared to peak transaction periods. Secondly, in contrast with temporary workers, part-time workers are frequently required to have firm specific knowledge of rules and procedures while temporary workers have non-industry-specific skills.

In the US temporary 'industry' close to a million workers are employed as temporaries at any one given time but, more significantly, at least 3 million people work as temporaries at some time during any given year. Average annual employment in the industry increased from 340 000 in 1978 to 944 000 in 1987. The temporary supply industry is growing at 3 times the growth rate of service industries and 8 times the rate of all non-agricultural industries (Carre, 1988). Between 1988 and 1995, the temporary help industry is projected to grow 5 per cent annually in comparison with a 1.3 per cent growth rate for all industries.

Occupationally, 52 per cent of Temporary Help Service Firm workers are employed in technical, sales and administrative support occupations, most of these in clerical occupations. The concentration of clerical workers is 2½ times their concentration in all industries. Sixty-six per cent of these workers work full-time. The second largest group is composed of operators, fabricators and labourers. These workers are more likely to be men; more likely to be Black and more likely to work part-time (Plewes, 1987; Carre, 1988). In addition, non-office temporary help appears to be growing faster than the clerical component of the industry. Agencies specializing in non-office temporary help accounted for only a third of the total temporary help service employment in 1972 but for 45 per cent of the total by 1982 (Abraham, 1988).

The temporary industry employs only a small portion of temporary workers. The largest portion are 'direct hires', employed as on-call workers in large firms and more and more frequently, in the public sector. The US federal government is one of the largest employers of temporary workers and under revised regulations can hire 'temps' up to four years without providing benefits or job security. Approximately 300 000 workers in the executive branch, including the postal service, are currently employed as temporary workers. Among the private firms with their own 'in-house' temporary labour services are Standard Oil, Hunt-Wesson, Beatrice Foods, Hewlett Packard and Atlantic Richfield (ARCO).

The role of the temporary industry and the temporary employee has changed over time. As the temporary agency has become more established as a labour market institution, more firms are restructuring work to use a 'permanent' temporary labour force to do certain jobs. Rather than a part-time phenomenon, temporary workers are more frequently employed on long assignments, for weeks and even months (Plewes, 1987). . . .

Who makes up the temporary workforce? Temporary workers are similar to the part-time workforce – they tend to be young and female. The best available information on the characteristics of this workforce, from the May 1985 Current Population Survey, indicates that 64 per cent of temporary workers are women and that one of three of them is between 16 and 24 years of age. Blacks are also over-represented in the temporary workforce. They constitute 20 per cent of the temporary workforce in comparison with 10 per cent of the workforce across all industries.

Temporary and part-time workers are overwhelmingly employed in large service and retail firms as a buffer workforce and to reduce labour cost. Their use has little to do with flexibility as it is conceived in the manufacturing model of flexible specialization but instead reflects the ability of service firms to rationalize and standardize their operations and to deploy work hours as needed. The other major form of flexible work, independent subcontracting, plays a more ambiguous role. In some cases independent subcontractors reduce costs for the firms buying their services. In other cases they provide a specialized service input.

Approximately 7.5 per cent of the employed in the US are *self-employed* in unincorporated businesses and an additional 2.6 per cent operate incorporated businesses. Businesses conducted in addition to full-time work are operated by between 2–3 per cent of the workforce and over 4 per cent of self-employed business owners own more than one business (US Department of Commerce, Current Population Survey, 1983). Changes in the role of self-employment workers in the American economy are closely related to the development of the service economy over the past 15 to 18 years. Self-employment declined steadily between 1950 and 1970, led by losses in retail trade. Self-employment in agriculture has always been high but continues to decline (from 67 per cent of all full-time equivalent employment in 1950 to 51.1 per cent in 1986). Despite continuing declines in sectors such as agriculture where self-employment has been the predominant employment form, self-employment began to grow after 1970, particularly in the service sector.

Self-employed independent contractors are an important source of high-skilled professional workers for industries needing short term specialized services. Independent contractors are prevalent in electronics, chemicals, and business services and among a set of professional occupations, including graphic design, engineering, technical writing, systems analysis and programming. These occupations have some common characteristics that make them amenable to independent contracting. They are highly skilled but their skills are not industry specific. Self-employed workers frequently work on projects that are non-routine and carried out within a definite time frame. The expanding use of self-employed workers is presumably related to the increasing project orientation of business, for example, the temporary employment of specialized teams to market a new product or to develop a specialized software application.

Self-employed independent contractors are also used to reduce both direct and indirect labour costs. This type of contracting, also subject to abuse of working conditions and 'off the books' payment, is typified by homework in electronics and apparel, but is also represented in services by clerical homeworkers. Clerical 'independent contractors' currently number between 5000 and 10 000 workers in the United States. There is reason to believe, however, that this form of work will expand. The companies which have homework programmes, including New York Telephone, American Express, Walgreens, Investors Diversified Services and Blue Cross-Blue Shield, are very large firms which hire large numbers of clerical workers. Of the approximately 250 companies with home-based work programmes, between 20 and 30 are known to be in the process of expanding their programmes (Appelbaum, 1985). The programmes now in operation are essentially pilot programmes which will be evaluated and redesigned to facilitate homework productivity and supervision. The attractions of home-based work for the firm are substantial. They include increased productivity, elimination of non-wage benefit costs, reduction of turnover, and reduced costs related to off-hours computer utilization as well as decreased office space needs. Only about 10 per cent of the home-based subcontractors are full-time workers.

As with other emerging forms of employment, there are significant differences in the self-employed workforce by sex and race. For example, among the self-employed, more women than men are sole proprietors (70 per cent to 60 per cent). More men than women own businesses in addition to full-time employment (17 per cent to 15 per cent) and more women than men own casual businesses. Casual businesses are defined in terms of their total earnings ($1224 average in 1983) but 37 per cent of 'casual' business owners reported working full-time at them. As a consequence of this distribution, self-employed women had annualized earnings of $3767 in 1983 while men's annualized earnings were $13 520 (higher than the $12 079 earned by female paid employees) (Haber, Lamas and Lichtenstein, 1987). Median earnings for male self-employed incorporated workers (according to 1987 CPS data) were $35 114 and for female, $16 669. For unincorporated self-employed workers, the medians are $17 942 for men and $7930 for women (Appelbaum and Albin, 1988). Despite these differentials, women are the fastest growing portion of the self-employed workforce. Between 1979 and 1983, the number of unincorporated self-employed women increased 5 times faster than men and more than 3 times faster than wage and salary women. More than 9 per cent of working women own a business in the US . . .

On virtually every measure, the US workforce stands out as being more 'flexible', that is more responsive to changes in supply and demand in the market, than its counterparts in other industrialized countries. Labour turnover, as measured in the percentage of job holders who hold jobs for two years or less, is higher in the US than in thirteen other industrialized countries with which it was compared in an OECD study (OECD, 1986, p. 51). Wage flexibility as measured in the dispersal of wages within sectors, is also greater in the United States than in industrialized countries generally (OECD, 1986). And, although the US has a lower percentage of workers in part-time jobs than the UK or Sweden, the share of total labour input (in terms of hours) by part-time workers is higher in the US. Part-time work is increasing faster in other industrialized countries but the US is unique with respect to the multiple forms and extent of flexibility that characterize its workforce. . . .

Some implications for the spatial division of labour

A politically-informed analysis of the type I have sketched in this paper can potentially affect the way we approach international comparisons of emerging patterns of work. It could, for example, encourage a second look at what appear to be patterns of convergence, say, in the proportion of part-time workers or in service sector employment. Although there has been a general trend across industrialized countries for part-time work to increase substantially and for this form of work to employ predominantly married women, there continue to be significant differences among countries in the age and gender composition of the part-time workforce: in how part-time workers are used: in how they are compensated: and in how the State regulates part-time work (de Neubourg, 1985; Standing, 1986). These differences are constituted through a political process, within the workforce and potential workforce, as well as between capital and labour.

If the allocation of work is politically determined, we cannot assume that services, including childcare, food preparation, and health care as well as business services, will be provided in the same way in all societies. In this regard, Urry makes the valuable point that empirical evidence does not support the notion that there is a natural trajectory toward ever greater levels of service employment in modern societies. The US, for example, has had a large service sector throughout the twentieth century than many European countries and in the second half of the century, it has grown at the expense of the extractive sector rather than manufacturing (Urry, 1987). So, the allocation of many types of service work in an individual society is an open question and a political question (Warde, 1986). Despite a general trend toward private provision throughout industrialized societies, we can expect that the allocation of work among home, State and private enterprise will take different forms.

A second set of implications for the spatial division of labour is drawn from the narrower sphere of industrial politics. Clark (1986) has demonstrated how the limits of union influence in the United States were constructed as a consequence of the geographic concentration of 'lead' industries, such as automobiles and steel and the consequent local community orientation of industrial politics. Decentralized bargaining was eventually codified in US labour law (Stone, 1981). The labour institutions constructed within this context may be incapable of responding to the needs of a workforce in flexible service jobs. The new forms of work and the new workforce are foreign to the major US labour unions which continue to operate in the 'mass-production' mode, for example, opposing part-time work rather than organizing part-time workers. The combination of a tradition of exclusionary practices with respect to women and minorities and a history of localized industrial politics makes it very difficult to regulate flexible work carried out in national firms composed of many small dispersed establishments. It is this 'legacy' which is, at least in part, responsible for the burgeoning of flexible forms of employment in the US.

The trends in US service industries also raise questions about the way geographers view processes which effect change in spatial patterns. The standardization and rationalization of services is eroding local differences in production and consumption. More people do much the same thing. Should this be of interest to geographers, who have traditionally been concerned with what makes places different from one another? The answer is yes, if only because underlying an interest in spatial difference is a more basic concern for the alteration of spatial patterns in and across places. In some service-dominant economies, the systemic rationalization of production and distribution in retail, health, insurance, and banking firms is altering local economies and employment possibilities to make them more similar from place to place. Thus, geographers need to examine not only those processes in production that create new forms of local difference but also processes the intent of which is to reduce local difference. Systemic rationalization intersects with both old and new sources of differentiation, including the place-based transformation of manufacturing production. It is in this emerging interaction that place is being redefined.

Conclusion

The introduction of two new elements could significantly alter the way we approach the emerging spatial division of labour. One is a concern for the role of labour in the transformation process and more particularly the legacy of mass-production trade unionism in shaping responses to the issues raised by new patterns of work, including 'flexible' work. The second is attention to changing forms of production in sectors other than manufacturing. If one looks at the organization of production in services, in which the majority of workers in industrialized countries are, in fact, employed, a significantly different set of patterns emerge than are captured in the manufacturing-oriented flexibility model. To understand new patterns of labour use, we need to examine how the organization of production in services differs from the manufacturing model. And, since service workers have worked, for the most part, outside the traditional wage–worktime bargain, we need to assess what effect that has had on tendencies toward flexibility in these sectors. Only an enlarged vision of economic change, one that incorporates an understanding of production in services and of industrial politics, will allow us to adequately interpret the emerging spatial division of labour.

Mrs Thatcher's vision of the 'new Britain' and the other sides of the 'Cambridge phenomenon'

P Crang and R L Martin, 1991 (*Society snd Space* 9: 91–116)

The other sides of the Cambridge phenomenon: 'south–south divides'

. . . To look at Cambridge as a whole has problematic implications for any use of it as a symbol of economic success. It raises the question of whether there are social and spatial divides within what in aggregate appears a prosperous and dynamic locality; whether there are segments of the local work force and population for whom the high-technology growth of recent years has brought few if any benefits. Thus in the same way that southern Britain is frequently equated with Thatcherite prosperity, the simple equation of Cambridge with 'success' may conceal as much as it reveals.

In fact, behind its outward appearance Cambridge contains many of the social and economic inequalities to be found more widely across the prosperous 'south' of Britain: it too displays elements of what has been identified as a 'south–south divide' (Murray, 1988; SEEDS, 1987). 'The South East is sometimes portrayed as a powerhouse for Britain's economic recovery, sometimes as a playground for the wealthy. Either way, it is argued, everyone in the South East enjoys a high standard of living. . . . Reality, however, is much more complex than this, and a good deal less comfortable or reassuring' (SEEDS, 1987, page 1).

Like other prosperous cities in the south of Britain (for the case of Swindon see Boddy *et al.* 1986), Cambridge amply illustrates this point, and we want to use it to sketch out, albeit somewhat schematically, how such divides might be geographically and socially conceptualized. To elaborate it we will focus on four aspects of the geographies of these divides within Cambridge. These are, first, the 'front' and 'back' regions (Goffman, 1959) that underlie the 'official versions' of the Cambridge economy; second, some of the areal segregations that are associated with these regions; third, the varied geographies of labour markets and individual 'work histories' that intersect in the city and which constitute part of the differential 'locales' of the city's inhabitants (or the space–time patterning of social behaviour: see Giddens 1984); and fourth, the 'overlapping' in Cambridge of very different growth service sectors (see Allen, 1988). In so doing, we hope to hint at some of conspicuous silences in the accounts we have looked at so far, especially in relation to the unemployed, the low paid and female employment.

If we examine the time–space paths of the numerous recent celebrity visits to the city by British and foreign politicians, as recorded in the local press over the past ten years, a clear division between 'front' and 'back' regions is readily apparent. The typical itinerary is a visit to the Science Park, followed by a reception at Trinity College, and often either a subsequent visit to one of the University science departments or, in the case of government ministers, an address to the University Conservative Association. Without doubt, the University, including the academic Colleges, the Science Park, and the adjacent Innovation Centre, are the 'front' regions, the dominant public realms. These are the areas of the city used by those who organise such visits to display Cambridge as a 'locality' to the outside world, and by those who visit them, from Margaret Thatcher to Neil Kinnock, to associate themselves with the politics such places are bound up in. These regions are promoted not only to notable visitors but also to the general visiting public, for example in the tourist literature we mentioned above (Ferguson *et al*, 1987).

But it is not only specific places that are celebrated or omitted in this way. Particular groups of people are obviously left out of the official versions, especially the unemployed, whose very existence is denied by the proclamation of 'full employment' to which we have already referred. Although the proportion of the work force involved is undeniably small, unemployment does exist in Cambridge. The manager of a local job-search programme commented to us that the notion of 'full employment' made his clients 'invisible people'. It is in fact not just a definitional issue but also a morally laden judgement on the relationships between capital and labour to declare full employment. Labour markets deal in a commodity like no other: there is no simple equivalence of jobs and people. The government's Employment Training Scheme has stressed this mismatch, and has used it as part of its policy campaign on unemployment and skills shortage, in the South of Britain in particular. The message is that there are in fact enough jobs, and that all that is needed is 'to train the workers without jobs to do the jobs without workers'. However, the problem is not simply one of training. We have already noted the unfilled vacancies in the Cambridge labour market, but there are cultural as well as skill barriers to the filling of these. In many consumer services, for example, customer contact staff are a part of the service product itself: their labour is not only a matter of training but also of promoting the image of the service or product. The young age-profiles of the waiting staff in the newer 'designer' restaurants and clothes stores in the city are not only because of the hours and the pay – though low pay is a massive barrier to anyone with a family to support – but also to what employers see as the need to present the 'right image' to their customers.

Likewise the cultural and social background of employees is also used by employers as a criterion of selection and discrimination: as one manager of a fashion retail store said to us, the residential origin of his employees is an important consideration: none came from the large council estates in the north of the city as they were not the 'right type of people' he was looking for. They lacked the 'cultural capital' (Bourdieu, 1984) to display and sell the middle-class clothes in the store. These cultural barriers are not limited to customer-contact jobs. They are also evident in the selection of unskilled ancillary staff by local high-technology firms, where employers are concerned, explicitly or implicitly, about 'organisational culture' (see Peters and Waterman, 1982). A common barrier in such cases is age; older employees are seen as problematic for a young dynamic company.

Thus, even if those seeking work have the requisite skills, various barriers still exist in the relations between job and employee. Nevertheless, the pressure is on in Cambridge, as in the south more generally, to eradicate unemployment by forcing the unemployed to take any job. The unemployment that remains in Cambridge (undoubtedly underestimated by official figures) includes two main types of people: those whom employers deem unsuitable, and those who are holding out for a job that uses their talents and pays a decent wage. Both groups refuse to be fitted into undesirable and low-paid jobs: and this reluctance to be forced into the bottom rungs of full employment, such as cleaning staff or kitchen work in restaurants, is then condemned as a preference to be voluntarily unemployed.

The display of certain regions of Cambridge in order to promote a particular image of the city obviously depends upon a selective areal segregation. Not all high-technology industry is located in the Science Park, but the Park is still able to function as a physical symbol for this type of industry. This is because of the spatial concentration there of such activities, because it embraces the 'greenfield' environmental qualities that are considered to be an essential part of the locational requirements of the high-technology sector, and because of the formative significance widely associated with the very concept of a 'science park' itself. Obviously, one needs to add a temporal dimension to these areal divisions; at night the Park and its prestigious buildings in effect become temporary 'back regions' as the low-paid, low-status cleaning staff move in. However, some 'back regions' also have a distinct physical spatial delineation and identity. In looking at the South East as a whole we have already distinguished, using the currently fashionable labels, 'post-Fordist' cities like Cambridge from 'Fordist' ones like Basildon. Yet within Cambridge itself there are neighbourhoods whose built environment and social relations are the products of the welfarism of the 'Fordist' era. The two large council estates of Arbury and Kings Hedges on the northern edge of the city are prime examples. With the Thatcher government's move against public housing and the social welfare system in general, such local areas have become increasingly marginalised. Of ten neighbourhoods in Cambridge officially designated by the city council as areas of welfare need, seven are highly dominated by council housing. The plight of these areas stems both from restrictions on council spending on the built environment of the estates, restrictions emanating from the central government's general fiscal controls on local authority spending, and the high proportion of their inhabitants who, because of low-paid employment or unemployment, are dependent on the government's increasingly restrictive welfare benefits (CCC, 1985a). These areas of Cambridge remain as a legacy of the pre-Thatcherite era, now ignored by those in the high-technology 'phenomenon' but unavoidably entwined with that phenomenon via the problems of growth we have discussed. . . .

Thus many of the research staff in the city's high-technology section, although often changing jobs within the local area, are initially drawn from well outside Cambridge, either from a national labour market or as students entering the University. In contrast, the majority of routine clerical, retail, and unskilled service jobs are held by indigenous people, born and brought up in the locality.

Such distinctions between different labour-market segments with different 'geographies' overlapping in Cambridge lead us to a more general point. SQP tend to see the growth of employment outside of high technology as a direct multiplier effect of the latter: 'because of high profits and high value added per employee, the firms generate above-average employment multiplier effects, especially given the historically low wage levels in the local economy' (SQP, 1985, page 35). However, although there have been some multiplier effects of this sort [Moore and Spires (1986) estimate a multiplier of around unity] there are pressures for growth other than from the high-technology sector. In a more general context, Allen (1988) has noted how two very different growth service sectors, namely financial and consumer services, actually have similar regional geographies in Britain, and thus 'overlap' in specific places. In Cambridge there are at least four such overlapping sectors.

First, there is a growing business, producer, and financial services sector, some parts of which are linked to and have been stimulated by the development of high-technology activity, but other parts of which serve expanding markets elsewhere in the country. Second, there has been a dramatic explosion in tourism over the past few years, with the annual number of visitors to the city currently exceeding three million (CCC, 1985b). This not only contributes directly to local income but also to local employment, for example in hotels, restaurants, and shops. Politically, the Thatcher government has pinned as much hope on the tourist industry as on the high-technology sector as a source of national wealth and jobs. Year-round tourism and the 'heritage business' is indeed an important component of the Cambridge economy. Hence the city could be seen as an exemplar of Thatcherism as much for this as for its high-technology. Third, there is the general boom in consumer services, especially retailing, involving new suburban superstores and a host of new fashion shops. The boom in this layer mirrors that which has occurred nationally, and has as much to do with the rise in real incomes, the expansion of credit, and the wave of consumerism and materialism that have characterised Thatcherite Britain, as with the multiplier effects of the city's high-technology sector. The current phase of retrenchment in the retail sector as the government's high interest rates begin to bite illustrates this. Fourth, over the past decade there has been a marked growth in 'educational business' in the city, particularly language schools for foreign students. These different service-sector layers overlap and interact locally to produce a highly segmented labour market . . . the vast majority of employment in Cambridge is outside the high-technology sector. Occupationally there are also marked contrasts. Table 1 gives the occupational breakdown of high-technology, office, retail, and leisure employment. As one would expect the last two are radically different, whereas the office sector has within it both clerical and professional employment, especially in financial services.

There exist, therefore, several other worlds of employment whose characteristics are largely ignored in the dominant accounts of high-technology Cambridge. These are not historical legacies, but themselves dynamic parts of contemporary Cambridge. Here we want to say a little on two such worlds, namely restaurant waiting and office cleaning, drawing on interviews and participant observation in each case, to illustrate some of the characteristics of the other sides of the phenomenon. One of the most striking features of most jobs in these activities is that they are very poorly paid: average earnings in

Table 1 Occupational breakdown of high-technology (high-tech.), office, retail, and leisure sectors, 1986–87.

Occupational groups	Percent of total employees			
	high-tech	office	retail	leisure
Management	10	17	12	18
Scientific, technical, other professional	47	42	13	9
Clerical	21	32	13	8
Skilled manual	11	5	12	9
Semiskilled and unskilled manual	11	1	12	11
Sales assistants	0	3	38	0
Waiting	0	0	0	32
General kitchen	0	0	0	13

Sources: CCC, 1986; 1987

some cases are as low as £2.00 per hour. Only recently have earnings begun to rise in response to labour shortages and competition from other sectors. The other striking feature is that these low-paid jobs frequently involve long and unattractive working hours, in order to provide flexibility for employers. Full-time kitchen staff and porters at one restaurant we observed usually work a minimum of seven full shifts, and average ten or eleven out of a possible fourteen (day and evening) shifts. They are paid £2.60 per hour plus 'tip-out' (a small share of waiting staffs' tips). The office cleaners we studied fared no better; most worked for minimal rates of pay at times before or after the normal working day for short, part-time shifts of one to two hours. Many also held poorly paid or unpaid day-jobs and some even held a third part-time job in the evening. Working multiple part-time jobs was not a matter of choice but necessary in order to build up a liveable weekly wage.

Examples of this sort make it only too clear that service employment in Cambridge is not limited to a 'postindustrial' technocracy. At the other extreme are the service jobs that are rooted in low pay and servility. Such jobs have a longer and perhaps more prominent history in Cambridge than in many other localities because of the traditions of employment within the various colleges of the University. Historically, the colleges have operated a low-paid but paternalistic system of employment for large numbers of 'bedders' (room cleaners), porters, and kitchen staff who constitute the world of 'backstairs Cambridge' (Barham, 1986). As major employers of this sort of labour, the colleges have undoubtedly contributed to the historically low-wage character of the unskilled sections of the Cambridge labour market, an influence in stark contrast to the University's role in stimulating high-paid, high-technology employment in the Science Park. These servile service jobs are themselves being restructured within the University but primarily via the growth of new forms of employment.

The new 'servile' jobs are in the expanding leisure and cleaning industries and they lack the paternalism and unpressured work atmosphere of the colleges. In particular, they have different sociospatial relations of consumption. For example, unlike 'bedding', office cleaning is frequently subcontracted out to other companies, to be done at times when the normal office staff are not there. The workers are distanced from those they serve, and they form a division of labour that is hidden, as figures who move round empty office blocks for an average of less than £3 per hour. Very different relations of consumption, and hence experiences of work, are found in other low-paid service jobs. Given the closer customer contact associated with restaurant waiting, social relations, and the experience of servility in this sector have been restructured to make them highly personalised rather than impersonal (Marshall, 1986; Whyte, 1948). In fact, this type of service employment often requires extensive 'emotional labour' (Hochschild, 1983), in that waiting staff are called upon by their employers to adopt an informal service relationship with customers, to be themselves, and to identify with and help promote the particular life-style image and 'consumer culture' (Featherstone, 1983) that the new 'designer' restaurants (such as Garfunkels, Browns, The Dome, and Old Orleans) seek to create.

Such worlds of work are far removed from that of the high-technology 'wunderkind' and the kinds of labour portrayed in the dominant account. SQP also have little to say about the distribution of individuals to these different worlds. In particular, they are silent about gender; they acknowledge it only once. When talking about female employment in business services they comment, in a patronising way that perhaps says as much about the authors' attitudes as about the specific context under discussion, that 'it seems that such women find it congenial to work alongside young technologists and professionals who have contemporarily liberal attitudes on the role of women in employment. Interestingly, however, we are aware of only a handful of cases in which women are among the principals in the high-technology companies themselves' (SQP, 1985, page 92).

What they have in fact encountered is the ambiguous position of women in the Cambridge labour market. The growth in employment in the city has, as elsewhere, seen a rise in women's participation, and not only in the least attractive jobs. Moreover, certain 'female' occupations, such as secretarial work, have seen pay rates rise because of labour shortages. The conditions for participation have also improved, as in the development of some creche facilities to attract women 'returning to work'. Yet this expansion of female employment is riddled with sexist ideologies. Women are concentrated in those service occupations that most prominently display forms of the servicing ethos. These include the worst-paid jobs such as cleaning, waiting, and other extensions of domestic roles. Waiting is an especially interesting example. Some local restaurants operate 'bans' on the employment of men as waiters. A common reason for such discrimination is what employers consider to be the preference of many customers (notably men) to be served by young women. Where waiting is 'unskilled' and routine then women employees are most numerous; on the other hand, where waiting takes on a skilled or 'silver-service' image, then men tend to predominate. Women are seen as having talents for being friendly and men are seen as being able to learn skills.

Within the high-technology sector itself a similar gender division is also discernible. As SQP note, there are but a handful of women working in private-sector research jobs in Cambridge. Women in high-technology companies tend to fill the personnel, administrative, clerical, and less often the marketing posts. Given the well-documented biases in education and society as a whole against women in science and technology (Cockburn, 1985), this segmentation is hardly surprising. The result is that in the world of high-technology, men are the central players at the 'cutting edge', and encounter women primarily as their clerical assistants or personnel officers (see Bowlby, 1990). Thus the rise of new forms of service employment has undoubtedly had advantages for women, advantages that 'market' feminists see as gains that labourism never provided in the past. The presence of women in many business services is testament to that. However, in Cambridge it seems that these gains have been associated with a particular ideology of gender relations, that of the servicing woman. Such an ideology not only affects the distribution of individuals to occupational places preconstituted by systems of patriarchy and employment discrimination (see Massey, 1988), but can also affect the nature of those places and the demands made upon them.

Conclusions: In what sense an exemplar?

In this paper we have outlined two tales about the same city that carry rather divergent interpretations of the nature of the Thatcherite vision of the 'new Britain' embodied in, or extrapolated from, the 'Cambridge phenomenon'. What has rapidly become the conventional or 'official' view of the 'phenomenon' is, in our opinion, a somewhat fragile symbolic vehicle, whether for the claims made for local high-technology development that abound in industrial geography, or for the claims made for the free-market enterprise economy that permeate the rhetoric of the Thatcher government and its supporters. In the first place, it is clear that the city's high-technology growth has brought its own problems and dilemmas, and is not altogether the success story it is often held up to be. Within political and academic circles Cambridge has assumed a key role in the construction of an 'ideology of high technology' in which it is implicitly assumed that high technology necessarily means high skills, high wages, job creation, and prosperity. Although if narrowly defined the 'Cambridge phenomenon' does have these characteristics, it is as misleading to view the local economy as a whole in these terms as it is to assume that the high-technology experience of Cambridge can, or should, be replicated elsewhere [what Sayer and Morgan (1986) refer to as the fallacy of composition]. The 'success' of Cambridge has been crucially dependent on a unique local combination of circumstances. Of course, as Keeble (1989) suggests, there may be aspects of the Cambridge experience that are of relevance for policymakers trying to develop high-technology and small-firm activity in other locations and regions, but such lessons are not, we would argue, all positive ones. As the case of Cambridge demonstrates, high-technology-based economic growth, even when clean and 'postindustrial', can pose problems of inflation, conservation, planning, and aesthetics as well as of congestion and labour-market imbalance. In this respect Cambridge highlights many of the adverse consequences of rapid growth also apparent elsewhere in the South East of Britain, problems which contrary to the government's belief are beyond the scope of free-market forces alone to resolve.

Our complaint, however, goes further. The official accounts of Cambridge's success focus their attention on a prominent though limited 'primary' segment of the city's labour market. There is, we have suggested, another side to Cambridge that indicates that the city is far from being a singular success story, that the development of a high-technology sector has not benefited all who live and work in the city. However, this alternative vision is not one that points to any simple social and spatial polarisation within Cambridge, even though examples of such polarisation do indeed exist, as the contrasts between the high-technology Science Park and the northern council-housing estates illustrate only too vividly. There are in fact several 'other sides' to Cambridge that tend to be dismissed or submerged in the dominant accounts: for example the unemployed, the low-paid, casual, and part-time employed, and women in and out of paid employment. Many of these groups form distinct 'secondary' segments of the labour market. Furthermore, we have argued that there are geographies to these divides that undermine the notion of a single 'locality' of Cambridge, and which are likewise ignored in discussions of the phenomenon.

Yet, ironically, not only do these other sides to the city's socioeconomy reveal the limited extent of the high-technology 'phenomenon', they simultaneously

117

exemplify some of the less celebrated but more pervasive features of Mrs Thatcher's 'new' Britain. Probably the most significant is the fact that in Cambridge, as in the nation generally, economic growth over the past decade has been characterised by a widening of social disparities. Cambridge, like much of southern Britain, has certainly witnessed a major expansion in employment; but not all of this has been high-waged and high-skilled, and as much has consisted of low-paid, unskilled jobs. It is this sort of work that has formed a major component of the government's proclaimed 'employment boom' and increased 'flexibility' of the labour market (Martin, 1986). But the quality and conditions of many of these jobs are often such as to raise real doubts about the true nature of the 'boom' and the meaning of 'flexible' working. Again, Cambridge illustrates the way in which in the context of this employment boom the unemployed, though now reduced in number, have become increasingly marginalised, not just economically and socially but also politically. And, as a city dominated by a variety of forms of service employment rather than just high technology, Cambridge exemplifies a Thatcherite Britain which is not reducible even in its successes to a 'postindustrial' utopia of 'enterprise individualism'. It is in far wider senses that Cambridge exemplifies many of the features of Mrs Thatcher's 'new' Britain.

CHAPTER 6

The organization of services in production

> [T]he large corporation provides much of the organisational tissue that binds places together. (Dicken and Thrift 1992: 288)

Chapter 5 highlighted the spatial concentration of professional, managerial and technical jobs in metropolitan-based control centres, and the more widespread dispersal of more routine and less-skilled service employment. This pattern has been especially influenced by two interrelated aspects of modern production:

1. The growing dominance of large organizations in service and other industries.
2. The use of computer and telecommunications technology by these firms, to help them grow beyond the traditional limitations of local markets.

This chapter explores these developments. It takes the view that large firms provide an important window into the operation of local economies. Not only are they the agencies through which economic changes are communicated to localities, but the variety of corporate responses to change, and their historical accumulation of activities at particular sites influences patterns of economic development. Large firms are also placed within a wider division of labour (Walker, 1989). The way organizations orchestrate their internal managerial, administrative and technical functions, and their external relations with suppliers are central to service location (see Chapter 5).

Organizational trends among services nevertheless involve more than simply the growth of large private and public bodies. These are increasingly subject to reorganization of their internal management and administration, and shifts in service provision between in-house departments and business service suppliers. The latter often favours small service businesses, which offer specialist expertise in a variety of areas. Networks of large and small supplier firms have evolved, allowing large firms to concentrate on their core business.

Information and communications technology has played a critical role in the growth of large firms and their recent reorganization (Hepworth, 1986, 1989). The success of large organizations depends on the key personnel who plan investments, organize production and sustain management arrangements to serve national and increasingly international markets. Information and communications technologies help these:

1. to interpret the flood of information which is the bane of modern business;
2. to develop successful new products and services in increasingly competitive markets;
3. to manage increasingly complex organizations which possess a myriad of links with suppliers of specialist expertise;
4. to communicate or trade in geographically dispersed markets, and to control their sites over long distances.

The growth of big business

Even up to the 1960s, most services were thought of as dominated by small firms, providing labour-intensive services to essentially local markets. This was true in retailing, personal services and even business services in specialized industrial and commercial regions. In the financial services, the advantages of control over large assets meant that in most countries outside the highly regulated US, banking was relatively concentrated into a few firms. However, mortgage provision and many branches of life insurance were less concentrated, reflecting the traditional dependence of firms on local savings markets. The growing role of the state in health, social security and education was associated with centralized, bureaucratic control, but actual service delivery was generally still a local matter, through relatively small units.

This atomistic pattern has been challenged by the growing influence of large service organizations, and by the influence of large organizations in other sectors on their business service suppliers (Howells, 1988). The growth of large, spatially dispersed organizations, relying on computer and telecommunications technology to transfer information over large distances within companies and to export business and even increasingly some personal services, encourages an uneven geography.

Consumer service markets, of course, remain geographically dispersed, and must be provided for at the local level. This means that small firms still dominate the numbers of private sector enterprises. Their share of sales has nevertheless generally declined for a range of products, including food retailing, electrical goods, clothing and footwear, as large competitors increasingly deliver services through extended distribution networks. The modern technical capacity to do this, combining information and transportation systems, depends on specialized expertise in retailing, or the franchising of personal services, such as fast food delivery, repair and maintenance work, or real estate transactions. Often, though the delivery of goods and services remain local, their availability, quality and cost depend on decisions made far away, in the headquarters of large organizations. Small organizations are thus no longer protected in many local consumer markets. To survive, they must specialize in the types of service they offer, including, for example, the hours they remain open. These service developments are evident in every high street or shopping mall, generally favouring larger centres, where national companies prefer to concentrate to ensure high business turnover (see Chapter 7).

The increasing tradability of business services

As a result of corporate growth and technical change, the traditionally dispersed geography of services has been overlain by highly uneven developments. This is particularly evident in business and information services. As we have seen, for much of the 1970s and 1980s this was the fastest growing employment sector in many advanced economies (Chapter 2), but it has also been the most spatially centralized in major metropolitan regions (Figs 6.1 and 6.2; Stanback and Noyelle, 1982; Gillespie and Green, 1987; Howells, 1988). In the regions of France, Germany and the UK, for example, the proportion of employment in business services ranges from 2 to 12 per cent, but only four regions have over 7 per cent. These are the Il-de-France, including Paris, the South East of England, including London, Hamburg in Germany and the Province–Alpes–Côte d'Azur region in France (Fig. 6.1). Similarly, in the case of the US, only 32 areas out of 183 had above the national average geographical concentration (i.e. location quotients >1.0) of producer services in 1985 (Fig. 6.2). Thus, paradoxically, though knowledge, ideas and opinions can now be circulated almost instantaneously by computer and telecommunications technology around the world, the personnel that interpret this information remain concentrated in a relatively few places.

Modern businesses face great uncertainty arising from global patterns of competition, high rates of technical change, new forms of capital availability, competition for investment and changing national and international laws and regulations. This has fed the demand of large organizations for information-based services. These are not only supplied by the business service sector. All firms provide some such expertise of their own, through their capacities to judge and plan commercial trends. Specialist business services have, therefore, grown to augment clients' in-house financial, production planning, marketing, information technology and human resource management functions. These two business service sectors, in-house and independent, increasingly interact and depend on each other (O'Farrell and Hitchens, 1990a,b; Wood, 1991a).

Thus, while much business service growth is intimately related to changes in information technology, this is not the whole story. It depends even more on escalating organizational demand for the service skills required to plan and exploit complex information networks. These

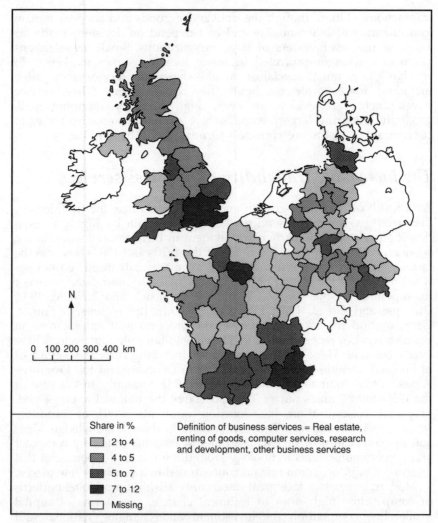

Figure 6.1 UK, France and West Germany share of business services in total employment, 1991 (*Source*: Labour Force survey)

may relate to the organization of production, including implementing strategic change, staff training or new forms of financial management, or marketing and product strategies. Business service growth is also affected by technical and organizational innovations in client sectors, including primary and manufacturing activities, the financial services, travel and tourism, the media, advertising or distribution. With rapid change, clients often need to seek outside expertise to complement their own skills, creating potential for new business services. The result is a complex pattern of shifts between in-house and specialist service production, involving new ways of managing client organizations in relation to their changing technical and market requirements.

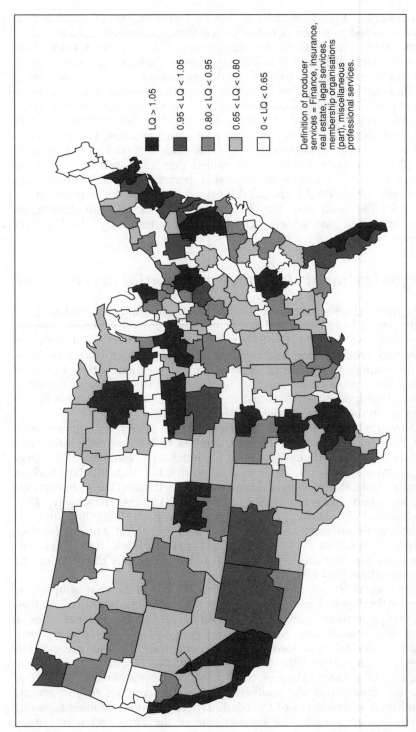

Figure 6.2. Producer services location quotients in the US, 1985 (*Source:* Beyers, 1989)

Generally, more decentralized, 'leaner' forms of business organization have emerged in recent decades, linked through networks of interaction with other organizations, including specialist business service firms (Coffey and Bailly, 1992). While these changes revolve around organizational and technological innovation, they are more fundamentally driven by attempts to utilize service expertise more effectively. This may sometimes favour small business service firms (Keeble, 1990; Wood *et al.*, 1993). Small firms benefit from modern low-cost communications and data processing facilities, to cover needs for which large organizations find it difficult to provide. Many such small service firms are based on skilled personnel 'spinning-off' from larger user organizations to exploit a perceived market opportunity (Howells, 1989). As the specialist business service sector expands, it too becomes the 'seed bed' for the development of new small firms, as founders seek to achieve greater independence by working for themselves.

Organizational structures and spatial development

Geographical research in recent years has thrown light on the locational strategies of large firms, and the ways the organization of their internal activities, and their links with external suppliers, affect the distribution of service work in local and regional economies. Much of this evidence is for manufacturing firms, contrasting the location patterns of their blue-collar production and white-collar service workers, including management, administration and technical staff. It is thus only indirectly relevant to the general analysis of service location.

Nevertheless, a predominant view of the effects of organizational structure on location has emerged, applied to both manufacturing and services. This emphasizes the hierarchical character of large organizations, which separate routine production of particular goods or services from supervision and management activities. These, in turn, are distinguished from top strategic control functions (Westaway, 1974; Goddard, 1979; Dicken and Lloyd, 1981). Each level has different locational requirements, creating distinct geographical patterns for each corporate function. The different qualities of work they offer also dominate the changing economic and social geography of the cities and regions where they concentrate.

Of course, the detailed implications of this somewhat simple and standardized model are highly complex. Nevertheless, in many countries, distinct patterns of headquarters, divisional and production activities, at least in manufacturing, show marked geographical similarities. Headquarters control functions tend to concentrate in and around key metropolitan centres (Box 6.1), while routine production has tended to disperse to smaller urban or rural areas, and even abroad. In the UK, this has meant that the southern part of the country, in and around London, is where many of the headquarters functions of manufacturing companies are found. The same is true of the Paris region in France,

---

Box 6.1: The concentration of corporate headquarters

As the need of the large manufacturing and service corporation for developmental and administrative functions has increased, these institutions have chosen to expand their head offices . . . in major urban centres. (Stanback et al., 1981)

The corporate headquarters is a good starting point for an analysis of the influence of business organization on service location. As large firms increased in size during much of the twentieth century, their corporate headquarters recruited ever-larger numbers of staff to administer, manage and support corporate networks. As head office demands for consultancy, marketing, legal, financial and computer expertise increased, their strong links with specialist suppliers encouraged the development of a complex of related business services (Hutton and Ley, 1987).

Despite some recent head office decentralization from the very largest urban centres, large firm headquarters continue to play an important centralizing role in the urban system. Head offices are attracted to large urban areas which offer 'ease of inter-organisational face to face contacts, business service availability and intermetropolitan accessibility' (Pred, 1977: 177). Howells (1988) demonstrates that large European companies are based overwhelmingly in major cities such as Paris, Milan, Frankfurt and Brussels. The UK has one of the most centralized corporate-urban systems in Europe. In 1987 some 271 of the head offices of the largest 500 companies in the UK were based in London, some 54 per cent of the total. London possessed 50 fewer head offices than in 1971, but the places that had benefited were largely in its hinterland, including, for example, Slough, Reading, Basingstoke and Bracknell. In contrast, the number of head offices from the 500 largest firms in the north and west of the country had declined from 132 in 1971 to 119 in 1987 (Marshall and Raybould, 1993).

The US urban system is less centralized (Semple, 1973; Stephens and Holly, 1981; Semple and Phipps 1982; Wheeler and Brown, 1985; Wheeler, 1985, 1986, 1988). For example, in 1979 only 132 (26 per cent) of the Fortune Industrial 500 companies were headquartered in New York, the main corporate base (Noyelle and Peace, 1991), and approximately 40 per cent in the North Eastern seaboard as a whole (Wheeler, 1990). By 1987 the numbers of head offices in New York had declined to 100 (25 per cent of the national total). The emergence of new dynamic companies in alternative centres in the south and west of the country (which together account for a third of large firm head offices; Wheeler, 1990); acquisition and merger activity and the decentralization of head offices from the centre of New York lie behind this trend (Wheeler and Brown, 1985). Nevertheless, even in the US, large metropolitan centres possess more than 95 per cent of the headquarters of the largest 500 companies (Wheeler, 1988, 1990).

Milan in Italy and Tokyo in Japan. In the US, these control functions are more dispersed, but still focus around the major metropolitan centres (see Parsons, 1972; Evans, 1973; Wheeler, 1985, 1987, 1988; Howells, 1988, 30–42; Schwartz, 1992).

The persistence of these patterns requires a significant augmentation of the corporate control model of location change. Something in the inherited character of places evidently influences the investment they attract in various levels of corporate activity. Some writers point to the skills available and inherited social attitudes towards different types of work (Massey, 1979b, 1984). Old industrial regions, with traditional manual skills and forms of worker organization, discourage the development of management functions, and become increasingly dependent on firms with headquarters elsewhere. Higher-order corporate functions, on the other hand, are focused in regions favoured by the technically and managerial skilled, with an internationally accessible, 'information-rich' business environment in which such skills can be effectively exploited (Goddard, 1979).

Coffey and Bailly (1992) suggest that large firm head offices, containing numbers of senior managerial and administrative staff, are major contributors to growth in metropolitan economies, not only because they support high-salaried, white collar work, but because of their extensive local linkages with professional support services. In turn, the concentration of high-level business services in similar types of location, in and around major cities, reflects and reinforces the wider pattern of corporate concentration. Intermediate control levels within corporations, perhaps based in divisional or regional offices, possess close ties with senior management, but also monitor routine production activities. They may therefore expand in second order cities with good access to both metropolitan and provincial industrial regions. Here they provide support for a regional base of business service suppliers (Dicken and Lloyd, 1981). Routine production sites, which are more dispersed, reflecting the costs of material processing, have few expert white collar staff, and thus also have few requirements for outside specialist services. The local development of service activities is thus 'truncated' (Britton, 1974; Marshall, 1979). The economies of areas of routine production are further weakened by lower incomes and consumer spending, with the absence of senior managerial or technical staff, and lower rates of new business formation, which tend to be initiated by such staff. The lack of corporate demand for services also reduces the availability of local support services, frequently provided by small business service firms to other local industries (Goddard, 1979).

This scenario clearly has general significance for the uneven geographical distribution of key business services (see Chapter 9). Patterns of trade are moulded by patterns of corporate behaviour. Howells (1988: 30–32) suggests that 'the geographical concentration of ownership and control' and the associated 'preference of headquarters units for large metropolitan areas' are common to both manufacturing and services, and that the 'branch plant economy' syndrome evident in peripheral regions in terms of manufacturing 'would appear . . . to be coming increasingly applicable in the service sector' (Howells, 1988: 36).

Howells is correct to point out the spatial concentration of the head offices of service firms (Box 6.1), but the relationship between the structures of large companies and their locational requirements differs between service and manufacturing industries. In the financial services, for example, corporate structures reflect the combination of international, corporate and consumer market requirements (Gentle *et al.*, 1991; Marshall *et al.*, 1992; Gentle, 1993). Professional services such as accountancy or legal services, where many firms operate as partnerships, have more devolved corporate structures (Leyshon *et al.*, 1988). In consumer services, such as retailing, the branch network of large firms is inherently more widespread than in manufacturing production. There are few administrative and support staff in the middle tier of such organizations, and head offices exert tighter control over local branches. Unlike those in manufacturing, these branches must also replicate a broadly uniform retail 'product' in each locality, not generally specializing in any aspect of the firm's service role (see the reading by Christopherson in Chapter 5).

Thus, large service firms respond to different market and competitive pressures compared with manufacturing, and to different technical requirements in production. The hierarchical structures of control appear generally simpler, although more devolved structures may sometimes require more varied forms of interaction of expertise. There is, however, a common requirement to spread functions over many localities, and for activities with higher levels of expertise to be spatially concentrated. The impact of large firm growth on employment in particular places thus still reflects corporate organization, and the way this is influenced by the existing character of places (Massey, 1984).

New corporate structures and changing urban development

Business management analysis since the 1970s has concluded, with increasing unanimity, that the old style hierarchical character of large firms is too rigid to respond effectively to changing business conditions in the late twentieth century. Now, firms must be able to react to rapidly changing patterns of competition based on new technologies, more varied customer requirements for goods and services, and the development of global markets (Brusco, 1982). Successful firms have introduced management and production systems which are less centralized and more responsive to market trends. These more 'flexible' organizations have also slimmed down their internal operations (Vonortas, 1990) to reduce costs and to allow companies to concentrate on what they do best. This requires new, cooperative forms of interaction with other firms, specializing in functions which might once have been undertaken in-house by the large firm. According to one commentator, this means that large firms are 'hollowing out' (Williams et al., 1990) as they subcontract more work to suppliers. To another, 'vertical integration . . . is coming apart under the pressures of the late 20th century' (Martin, 1990). It is also argued that greater interfirm collaboration and the need for speed of response promotes higher degrees of responsibility throughout the corporate network, because large firms need to be able to react quickly everywhere at once (Drucker, 1988). Thus, large firms become 'a collection of local businesses rather than a corporation' (Taylor, 1991: 96).

These corporate changes may thus be replacing the hierarchically based spatial division of labour which was underpinned by the older corporate geography discussed above (Storper and Christopherson, 1987).

127

Capital flexibility and spatial clustering may be becoming more dominant, although the rate of this change and the extent of its impact has been a matter of active debate (Schoenberger, 1987, 1989; Gertler, 1988, 1992; Lovering, 1990). In rapidly changing markets, flexibility in production and cost control requires firms to turn to subcontracting. As this develops, contractors themselves come increasingly to rely on external suppliers, creating further disintegration. So, with a 'progressive externalisation of production' (Scott, 1992: 13), it is claimed 'selected sets of producers... tend to converge around their own geographical centre of gravity' (Scott, 1988a: 177).

The **reading** by Storper and Christopherson (pages 134–43), which describes how these 'powerful agglomeration tendencies' operate in the Hollywood motion picture industry, is part of a growing literature on the development of new industrial districts (Chapter 4, page 65). They show how a shift from the large-scale production of pictures in big studios, to the contracting out of movie production to smaller independent studios, led the industry to be re-concentrated in Los Angeles. This is explained in terms of the need to minimize the extensive transaction costs associated with contracting out, and for specialist labour to fill an array of tasks associated with more flexible production.

Like much of the literature on industrial districts, their argument is couched in terms of the organization of manufacturing industry, focusing on the making of films, and treating services as part of the process of material production. In fact, movie-makers are increasingly part of large entertainment companies which may manufacture videos, TVs and film equipment, but also own cinemas, TV stations and cable companies, and may be involved in newspaper or book publishing. Here control of the distribution channels for films, and their marketing across multiple media, has become as important for commercial success as simple efficiency in their production. Thus, the spatial concentration of film production in Hollywood and the growth of small firms is but one part of a global entertainment industry where traditional boundaries between cinema, TV and entertainment are becoming increasingly blurred by technological changes and the activities of large conglomerates (Aksoy and Robins, 1992). Even within movie production itself, the changes Storper and Christopherson describe are clearly driven by services, and by the reorganization of the expertise characteristic of service functions (pages 140–3). The value of the 'product' depends more on the managerial, artistic and technical skills focused in Los Angeles, than the lower costs of material production there.

In a very different context, drawing on the insights of Storper and Christopherson, Coffey and Bailly argue that the growth in contracting out by corporate head offices lies behind the expansion of financial and business service firms in metropolitan areas. As firms increasingly rely on external contractors, the linkages between head offices and business service suppliers 'are becoming even stronger and more self-reinforcing than in the past' (1992: 865). The growing complex of corporate service activities in metropolitan areas, according to Lipietz, could in turn provide the impulse for a return, 'not only of factories and offices to urban zones, but also a new quantitative growth of metropolises' (Lipietz, 1993: 13).

There is clear evidence for significant change in dominant corporate cultures and structures. This is again more complex than the above description suggests, however, and more ambiguous in its impact on cities. Trends towards the contracting out of services by head offices in large firms are being complemented by others towards internalization of functions, especially at manufacturing sites (Marshall, 1989; O'Farrell *et al.*, 1993). Head offices are not only contracting out more business services (Aksoy and Marshall, 1992), but also slimming-down and decentralizing their functions, undermining the economic base of metropolitan centres. Thus while the growing network of links between firms and their service suppliers may favour large cities as centres for strategic management, the cities are becoming less important for routine administrative work, which is being decentralized to smaller centres frequently in their hinterlands (Box 6.2).

Box 6.2: Does corporate reorganization mean new patterns of metropolitan development?

Given the evidence for corporate reorganization, how might the character of head offices have changed in the last decade or so, and what implications do any changes have for metropolitan development? Somewhat surprisingly, there has been no large-scale survey of corporate headquarters since Crum and Gudgin's (1977) UK analysis. However, evidence from a small survey of 20 British-based corporate headquarters (Aksoy and Marshall, 1992; Marshall and Raybould, 1993), and various examples reported in both the management literature and the press (Vontoras, 1990; Taylor, 1991; *Guardian*, 1993), suggest head office employment has declined sharply in the last decade, and many head offices no longer employ large numbers of white collar staff. In the Aksoy and Marshall survey, for example, though the firms were all large multinational companies, only one head office had more than 300 employees, and 14 of the 20 head offices had been subject to substantial rationalization during the last decade.

In some firms the decline in head office employment simply reflects an attempt to reduce the indirect costs associated with running the business. Head offices are prone to such rationalization simply because they often employ large numbers of highly paid staff at expensive sites in large cities. But for many companies, headquarters rationalization is a result of a *decentralization of administrative responsibility* to lower levels of the corporate hierarchy. This reduces costs by making operating units more aware of the indirect costs associated with running their businesses. More importantly, however, such units are, in many cases, more informed about the day-to-day needs of their business than a remote head office, so delegation of responsibility increases the speed with which the company can respond effectively to change.

A related trend is an *increase in the proportion of services procured by head offices from external providers*. As the range and amount of support services required by firms have grown, new services are increasingly purchased externally. This allows firms to retain a 'lean' management structure, while satisfying their demand for new services. In addition, in response to competitive pressures, firms use external suppliers to reduce costs. Firms also believe that external suppliers provide greater choice of services and access to new ideas without internal fixed investment to worry about. The services most likely to be contracted out are non-strategic in nature, such as cleaning, catering, legal, computer or property services.   (cont.)

Box 6.2 Does corporate reorganization mean new patterns of metropolitan development? (continued)

   Though the decentralization of corporate decision taking and the contracting out of service functions inevitably lead to a reduction in the range of functions carried out at the head office, its power has not declined. Head office control of operating units is still maintained through tight financial monitoring. Head offices have simply specialized, becoming financial control centres, and bases for strategic manage-ment and planning.
   These changes in corporate head offices have important implications for metropolitan economies. Head offices in large companies, by contracting out their growing demand for support functions to local suppliers, could contribute to the further expansion of service industries in metropolitan areas. This certainly seems to be the view of Coffey and Bailly (1992) who envisage corporate head offices acting as a 'honey pot' for a range of local business service suppliers. They also highlight increasing tendencies for head offices and their suppliers to co-locate as contracting out increases. On the other hand, this optimistic scenario needs to be set against the evidence that competitive pressures on large firms and subsequent rationalization and reorganization may reduce the commitment of large firms to metro-politan areas. Headquarters employment decline, and related decentralization of head office administration to lower-cost areas, could help undermine the metropolitan economy.
   On balance, it seems likely that as a result of corporate reorganization, a more fine-grained distribution of economic activities is developing. The central area in the largest cities are increasingly becoming the base for corporate command functions and related high-quality business and financial services, while the hinterland provides the bulk of the administrative support functions (Wood *et al.*, 1993). The decentralization of corporate headquarters and administrative functions from the largest metropolitan centres (see Box 6.1), together with the emergence of new fast growth companies provides the basis for alternative service centres, but businesses still rely on the largest metropolitan area for the highest-quality services (Schwartz, 1992).

The impact of technological change on corporate structures

The growing integration of computer power and advanced tele-communications has significant implications for corporate structures. In the second **reading** Hepworth focuses on the take-up of such technologies by large firms, the main users of information and commun-ications technology (pages 143–54). He argues that such technologies are critical to the competitiveness of large businesses. But his analysis of the impact of the technology is not prescriptive. This is consistent with our view that developments in micro-electronics-based technology provide firms with new opportunities to utilize and develop their service skills and expertise, rather than determining particular forms of outcome (Chapters 4 and 5, pages 72–5 and 98–102). Hepworth's case studies illustrate a range of organizational impacts. They present a 'complex picture of organisational and spatial tendencies, rather than centralisation or decentralisation and vertical integration or disintegration' (Hepworth, 1989: 97).

Hepworth shows how new technology makes an *important contribution to shifts in organizational boundaries,* described in the last section.

1. Innovations in new technology lead to the development of new services, which non-specialist users can obtain from their service suppliers. Information and communications technology also allow non-specialist firms to carry out some new service functions themselves, e.g. to print their own advertising material. However, at the same time, the technology requires new specialist advisory, training and maintenance services, from service suppliers, to make the best use of it.
2. The case studies also demonstrate efficiency gains resulting from the use of computer networks. This leads to a reduction of internal bureaucracy within large firms, which is central to the growth of flatter, less hierarchical 'information-based' organizations described by management specialists such as Drucker (1988). 'Lean' internal operations may be associated with the use of external suppliers, to meet intermittent demands for services. Improvements in computer networks could also increase the capacity of the firm to handle outside suppliers, leading to the development of a 'network firm'. On the other hand, improvements in internal efficiency could in theory facilitate greater internalization of service functions, where the user firm becomes more efficient than service suppliers.
3. Information and communications technologies also affect the relationship between large firms and their service suppliers. In related work, Gillespie and Robins (1991) suggest that deregulation of telecommunications enables large users to by-pass public networks, to which all users have access, developing their computer services on private lines. As a result, a complex patch-work of limited access private networks is developing (Chapter 8, page 188). These networks increasingly tie companies and their suppliers together, and the latter may become dependent on the large firms controlling access to the network. Furthermore, where small- and medium-sized firms continue to depend on the public network, they may have access only to lower grade facilities (Gillespie and Williams, 1988).

Hepworth also shows how new information and computer technologies *expand the geographical reach and increase the locational flexibility of large firms.*

1. They release service firms from the requirement that production and consumption take place in the same location.
2. Computer networks may reduce the minimum scale of efficient production operations, allowing branches to be set up in new market areas.
3. Improving communications increases the 'control space' of large firms, again facilitating the geographical expansion of the firm.

4.   Multilocational firms may be able to locate functions in a more cost-effective manner, with communications technology opening up many new patterns which were not possible before. Hepworth's case studies suggest that the widespread use of information and communications technologies within large companies allows production and routine administration to become more widely dispersed, while activities concerned with critical aspects of management and control remain tightly centralized (see also Marshall, 1984; Hepworth, 1987, 1989). This is consistent with the trend towards delegation of responsibility within organizations discussed above. Operational responsibility may be decentralized within a framework of strict central financial monitoring and control, and centralized strategic planning, both of which are assisted by computer information and control systems (Marshall and Raybould, 1993).

Finally, although ties between firms and their suppliers of services in the local area may be growing (see the reading by Storper and Christopherson), developments in information and communications technology are increasing the connections between firms in widely separate urban centres. Microelectronics-based technology draws cities and the firms in them into global urban systems. Although some services are developing on the basis of intense local ties, the economies they support are increasingly integrated across the globe (Boxes 5.4 and 9.1).

Further discussion and study

Guided reading

Aksoy A, Robins K 1992 Hollywood for the 21st Century: global competition for critical mass in image markets. *Cambridge Journal of Economics* 6: 1–22

Coffey W, Bailly M 1992 Producer services and systems of flexible production. *Urban Studies* 29: 857–68

Hepworth M 1989 *Geography of the information economy.* Chapter 5 Belhaven, London

Marshall J N et al. 1988 *Services and uneven development.* Oxford University Press, Oxford: Chapter 6

Activity

1.   Examine a business directory (e.g. Yellow Pages, Kelly's, Kompass, Standard or Poors) for any sizable town or city, or part of the central business district of a large city.
     Enumerate the numbers of firms in various services which primarily serve other businesses, rather than consumers. What sectors are well developed, or under-developed? Which firms locate together, and why do you think this occurs?

What companies offer (a) advice on the implementation of modern computer and communications technology and (b) modern communications services (for information transmission, verbal communication, rapid parcel delivery, etc.).

Is it possible to identify firms that have their headquarters there? If so, estimate how many there are, and, if possible, their local and total employment.

Write a short 'consultants report' on the status of the business services for the local agency responsible for economic planning.

Consider

1. In 1986, Hepworth declared that, 'innovations in information technology will transform urban and regional systems', but also that 'networks are essentially instruments of organisational control'. From the case studies he cites in the reading, summarize the main implications of these statements for current urban and regional development.

    In the light of this analysis, how do you think a region might gain a 'comparative advantage' in trade in network-based information services?

2. What is meant by 'flexible specialization' in the reading by Storper and Christopherson? Why has it caused the reconcentration of the motion picture industry into Los Angeles?

    What predominant pressures have undermined the large, vertically integrated production firms: changing production *costs*; the impact of new *technology*; changes in government *regulation*; changing sources of *finance*; growing *competition*; their declining control of key *expertise*; greater *uncertainty* of success in film making; greater *trust* between independent agents; growing *union* power; the development of new *markets*; or what else?

3. To what extent is the picture of vertical disintegration and agglomeration depicted by Storper and Christopherson consistent with the case studies of firms using information and communications technology presented by Hepworth?

4. Read the article by Aksoy and Robins on the movie industry. What differences are there in their picture of industrial reorganization compared with the Storper and Christopherson reading? To what extent do Aksoy and Robins undermine the view expressed by Storper and Christopherson that flexible specialization, vertical disintegration and local agglomeration are significant features in the movie industry?

Discuss

1. How far does the greater availability of information create rather than solve problems for modern organizations? How is the answer to this question reflected in the ways firms operate?

2. Why should one organization get another to provide services for it,

rather than employing their own staff to do it for them? Focus on particular types of organization (e.g. manufacturing firm; banking corporation; government agency; retail chain; charity).

Why, in some cases, might the 'client' firm decide later to do the work itself?

3. Why have business services been growing so rapidly in recent years? What geographical impacts stem from this growth? To what extent have large organizations influenced geographical patterns of business service location?

Readings

Flexible specialization and regional industrial agglomerations: the case of the US motion picture industry

M Storper and S Christopherson, 1987 (*Annals of the Association of the American Geographers* 77: 104–17)

Despite its glamorous image, motion picture production was for much of its history a routinized factory-like process, under the control of large, vertically integrated firms. In the 1950s, however, all this began to change, and by the late 1970s the industry had been dramatically transformed. In the contemporary motion picture industry most films are made by independent production companies, which subcontract work to small specialized firms. This type of production organization can be described as 'flexibly specialized'.

The trend toward flexible specialization can be observed across a range of industries, extending even to classical mass production industries like automobiles and steel. Piore and Sabel (1984) suggest that this trend contradicts prevalent assumptions that mass production is the culmination of sectoral development. It also raises questions about location theories that extend this assumption about sectoral development to the spatial trajectories of industrial sectors. In this study we first review the theory that underlies subcontracting and flexible specialization and then present evidence on flexible specialization in the motion picture industry. . . .

Our findings show that over the past two decades motion picture employment and establishments have reconcentrated in Los Angeles even as most filming of motion pictures has moved outside the region. If the reasons for this concentration apply generally to flexibly specialized industries, and we suggest that they do, then flexible specialization can have a dramatic impact on urbanization, regional development, and trade patterns.

Flexible specialization as a form of industrial organization

Since about 1930, US and Western European literature on industrial organization has been dominated by the idea that a 'mature' industry is one in which large firms carry out commodity production with mass production methods (Burns 1934; Kuznets 1930; Schumpeter 1942; Vernon 1966; Storper 1985). Underlying this conception are models of sectoral development that posit as a 'normal' industrial development path one that starts with small-scale production and a disintegrated organizational structure and ends with a mass production system of vertically and horizontally integrated firms. . . .

Several years ago, however, this conception began to be challenged. Berger and Piore (1980), for example, demonstrated that small-scale production by subcontractors is an enduring correlate to mass production. . . .

More recently, Piore and Sabel (1984) have argued that the paradigm of mass production is based on selective readings of both industrial history and the industrial present. They claim that the particularities of twentieth century industrialization in the US and Western Europe cannot be transformed into a universal developmental logic and that mass production was only one of several possible forms of industrial organization that could have evolved in the early twentieth century. One contemporary example of sectoral development that contrasts with the dominant view appears in the Tuscany and Emilia-Romagna regions of central Italy. The firms of 'the Third Italy' produce specialty goods, in industries including textiles and automobiles, under short production runs and using skilled workers (Bagnasco 1977). They are small, flexible firms, capable of responding quickly to changing market conditions and connected to the market via contracts with other firms (Brusco and Sabel 1981).

In such flexibly specialized systems production is organized around the interactions of a network of small firms. These small firms specialize in batch or custom production of general classes of outputs, whereas, mass production firms are committed to the production of specific outputs in large quantities. The production system as a whole is flexible because each production project can be organized with a different mix of specialized input-providing firms. In more conventional parlance these firms are subcontractors in a system of production that is vertically disintegrated. The system is what Bagnasco (1977) calls a *fabbrica diffusa*, a regional production complex of firms interconnected by their numerous transactions with each other.

The organizational history of the motion picture industry

The motion picture industry was not always vertically disintegrated and flexibly specialized. In the 'golden age of Hollywood,' motion picture production resembled that of large-scale manufacturing industries with routinized production processes. From 1920 to approximately 1950, the motion picture industry was a concentrated oligopoly. Seven major studios owned their own theater chains, and in major regional markets of the US (cities with more than 100,000 residents), five of these firms controlled 70 per cent of first-run theater capacity. In 46 per cent of all markets, one of the major studios controlled distribution to all theaters (Waterman 1982). With market outlets assured, the studios could standardize the product and routinize production. They secured actors on long-term contracts and set them to work on a schedule of 20 to 40 films per year. Production groups, required to complete a film every five to seven working days, were backed up by permanent staffs of pre- and post-production workers. Filming was organized via the 'continuity script' so as to shoot similar scenes together rather than in their chronological order in the story.

This period of production rationalization was followed by one of internal reorganization, brought on by the forced divestiture of the studios' theater chains in federal antitrust action in 1948 and then by television in the 1950s (Christopherson and Storper 1986). From 1950 to the end of the 1960s the major studios dominated production within a hybrid structure that included independent producers and a significantly reduced in-house production schedule. During this period, the major studios responded to the changing profit environment by differentiating the film product and expanding the market for theatrical films.

After the recession of the early 1970s, a distinct industrial organization began to crystallize. The major studios maintained a firm grip on the financing and distribution of high-budget theatrical releases and moved into production for television. They also reduced their permanent workforces and in some cases sold production facilities in order to reduce fixed capital stock. In the contemporary motion picture industry, production firms specialize in generic tasks, subcontracting their services and equipment to a producer who organizes the film project. Production is carried out through independent production companies, either with or without major studio affiliation. Their scale of output is limited and the scope of their activities is relatively narrow. The production process is no longer carried out within the firm but instead has moved to the external market, carried out through a series of transactions linking firms and individuals in production projects.

Moreover, the technological revolution in entertainment has dissolved the boundaries that previously separated film, television, music, and video products. Specialized establishments can now reduce their risks by marketing their services across industries. In this entertainment industry complex, specialized production firms are combined and recombined as they work on various projects. A hierarchy of control remains, however. The major studios continue to dominate financing and distribution, retaining effective control over product definition and marketing.

A profile of vertical disintegration in the motion picture industry

In an industry where vertical disintegration is occurring, the role of large firms in direct production diminishes over time. Instead, production is carried out by a large number and a more diverse set of firms. We constructed a file of films whose production organization and location were reported in the 'film charts' of two major industry publications, *Daily Variety* and *The Hollywood Reporter*. The sample was taken at five-year intervals between 1960 and 1980, with a supplement for 1984. This resulted in a sample of approximately 1,200 films.

Table 1 reports what this sample reveals of change in the types of production organizations in the motion picture industry. There is a clear trend toward production by independent companies and away from the major studios. Over the study period 43 per cent of our sample films were produced by independent production companies, frequently established to produce just one film project and then dissolved. The major studios (Paramount, Universal, Twentieth-Century Fox, MGM, Warner Brothers) produced 47 per cent, and the mini-majors (Disney, Cannon, Orion) produced 9 per cent. The remaining 1 per cent was produced by television production companies. More importantly, the proportion of motion pictures produced over the period by different types of production organizations has shifted quite dramatically. In 1960, less than one-third of the pictures were produced by independents (alone or with others), whereas two-thirds were produced by majors or by majors along with another producer. By 1970, independent production had begun to rise steadily. In 1980, the independents continued to produce more films, even as the total number of films produced declined. These figures suggest that the major studios have shifted a considerable proportion of the managerial responsibility for motion picture production to independent producers.

The studios' increasing tendency to disintegrate managerial responsibility for production is further suggested by an analysis of the date of founding and pattern of employment in production houses. Our sample of 231 such firms includes those specializing in the production of commercials, industrials, animation, and special effects as well as in theatrical and television production. The temporal pattern of establishment of these firms replicates other data we have indicating industry cycles. Most of the firms were established after 1970. Of 200 firms reporting establishment dates, 140 were established between 1971 and 1982 and only 39 between 1961 and 1970. The largest surge in firm establishment was in the late 1970s . . .

To discover the nature and extent of the vertical disintegration of production inputs, we examined the births and deaths of selected film-making services and facilities in the Los Angeles area . . .

The number of facilities and service firms all increased during the period. Given that motion picture output was stable for the index years, the magnitude of these increases provides convincing evidence of a trend toward vertical disintegration. Certain production activities were spun off from the large integrated studios into increasingly specialized establishments. The greatest gains in number of firms between 1966 and 1981 were among firms whose functions directly relate to the production process – production companies, rental studios, properties, editing, and lighting. In every case, the major leap in the number of firms was between 1974 and 1981, after the cyclical downturn of the early 1970s.

Table 1. Number and per cent of productions by organization type, selected years, 1960–80

	1960	1965	1970	1975	1980	Total
Independent	42(28)[a]	40(21)	93(45)	138(57)	129(58)	442(43)
Major	100(66)	130(68)	96(46)	81(33)	69(31)	476(47)
Mini-major	9 (6)	20(11)	18 (9)	24(10)	24(11)	95 (9)
Totals	151	190	207	243	222	1,013

Source: Daily Variety for years given.
[a]Per cent of production in given year appears in parentheses

The location of the motion picture industry

The location of employment and establishments Table 2 shows changes in the location of employment in the motion picture industry. SIC 7813 (motion picture production) clearly concentrated in California between 1968 and 1981, with a temporary drop in California's share in 1974. Between 1968 and 1974 employment in SIC 7814 (television) became more concentrated in California; though some of this gain was lost after 1974, California's share of total employment increased from 1968 to 1981. Given the strong role that New York has played in television production from its inception, this shift indicates a decisive change in the locational logic of television production. In SIC 7819 (services allied to motion picture and television production), employment has always been highly concentrated in New York and California, but recently California's share has increased at the expense of New York's.

Table 3 shows the location of business establishments in the industry. Establishments in the motion picture industry proper (SIC 7813) show a tendency to decentralize nationally. New York's position as secondary center is giving way to a more dispersed pattern, while California's position as the principal national center remains stable. In allied services and television, by contrast, California's share increased significantly between 1968 and 1981. Once again, the relative positions of New York and California in television production shift decisively. In 1968, New York and California had almost the same number of establishments, but by 1981, California had twice as many as New York.

Table 2. Location of employment

State	1968		1974		1981	
	Total	% of US	Total	% of US	Total	% of US
SIC 7813: Motion pictures						
California	10,252	66.3	5,490	59.3	18,807	73.4
Florida	193	1.3	157	1.1	148	.6
Texas	202	1.3	417	2.9	368	1.4
New York	2,611	16.9	2,173	15.2	1,870	7.3
Illinois	712	4.6	[500–999][a]	–[b]	1,139	4.4
New Jersey	–	–	88	.6	51	.2
US	15,461	100.0	14,319	100.0	25,611	100.0
SIC 7814: Television						
California	5,197	50.0	14,839	72.9	21,626	63.2
Florida	160	1.5	[20–99]	–	302	.9
Texas	193	1.8	[20–99]	–	[250–499]	–
New York	4,076	39.2	2,423	16.8	6,755	19.7
Illinois	178	1.7	798	3.9	397	1.2
New Jersey	–	–	–	–	130	.4
US	10,388	100.0	20,359	100.0	34,226	100.0
SIC 7819: Allied services						
California	–	–	9,663	63.3	12,205	67.2
Florida	–	–	206	1.3	73	.4
Texas	–	–	131	.9	437	2.4
New York	–	–	3,110	20.4	3,135	17.3
Illinois	–	–	382	2.5	426	2.3
New Jersey	–	–	90	.6	85	.3
US	–	–	15,274	100.0	18,169	100.0

Source: US Department of Commerce.
[a]Figures in brackets report only range. No percentage calculations are made for these data.
[b]Indicates no data reported or percentage cannot be calculated.

During the study period establishments in SIC 7814 and SIC 7819 became more spatially concentrated than those in SIC 7813. The discrepancy is probably due to the much larger establishment size in SIC 7813 (the movie industry proper), where the lowest proportion of establishments but the highest proportion of employment (more than 73 per cent) is concentrated in Los Angeles. The largest studio properties remain in Los Angeles, which has more than 220 facilities exceeding 20,000 sq. ft., as compared to fewer than 20 in this size class in the nearest competitor, New York. Because of the abundance of studio space in the Los Angeles region, the construction of small facilities has not been economically feasible. The major studio facilities (which are complexes of many stages in one location) are a residue of the earlier studio system, and the major firms now rent out these facilities for independent efforts as well as using them for their own productions.

Most of the establishments in SIC 7813 (motion pictures) outside California are small companies or stages that do not account for a large proportion of filming. We do not have figures on the physical size of these establishments, but we estimate that a much larger share of production capacity is concentrated in Los Angeles than our figures on establishments allow. Moreover, many of the

Table 3. Location of establishments

State	1968		1974		1981	
	Total	% of US	Total	% of US	Total	% of US
SIC 7813: Motion pictures						
California	263	39.5	534	41.9	414	40.5
Florida	12	1.8	33	2.6	19	1.9
Texas	14	2.1	36	2.9	25	2.4
New York	199	29.9	303	23.7	214	20.9
Illinois	18	2.7	53	4.1	57	5.6
New Jersey	–[a]	–	24	1.9	17	1.7
US	666	100.0	1,279	100.0	1,023	100.0
SIC 7814: Television						
California	198	40.4	409	41.9	607	48.4
Florida	12	2.5	12	1.2	25	1.8
Texas	10	2.0	17	1.7	44	3.1
New York	175	35.9	306	31.3	311	21.9
Illinois	20	4.1	46	4.7	39	2.7
New Jersey	–	–	–	–	21	1.5
U.S.	490	100.0	978	100.0	1,420	100.0
SIC 7819: Allied services						
California	–	–	284	39.7	536	49.8
Florida	–	–	14	1.9	11	1.0
Texas	–	–	13	1.8	28	2.5
New York	–	–	233	32.5	289	26.5
Illinois	–	–	36	5.0	33	3.1
New Jersey	–	–	12	1.7	21	1.9
U.S.	–	–	716	100.0	1,077	100.0

Source: US Department of Commerce.
[a]No data reported or percentage cannot be calculated.

facilities outside Los Angeles are double-counted, because they are adjuncts to regional television or commercial production facilities that can be pressed into service for the occasional motion picture project. Some are rental studios built on speculation and used infrequently. The exceptions are the studio complexes of Earl Owensby and Dino de Laurentiis in North Carolina and of Burt Reynolds in Georgia, and the smaller rental facilities in New York City. In effect, the spatial spread of establishments in SIC 7813 is more apparent than real.

The location of filming Filming is more widely dispersed than is the location of employment and establishments (Table 4). The peak year for filming feature films and made-for-television films in metropolitan Los Angeles was 1960. The trough in both relative and absolute terms was 1970. Since 1970, the shares of both Los Angeles and California as a whole have risen but much more slowly than the concentration of employment and establishments; this implies that Los Angeles is recovering from a steep decline in its share of filming activity in the 1960s but will not reassert its position as the undisputed center of filming activity in the US.

One of the most prominent aspects of disintegration has been the rise of independent production companies. These different types of production companies show marked differences in the location of filming. Of the films shot in California but outside Los Angeles between 1965 and 1980, 68 per cent were made by independents and 29 per cent by major studios. Of those made in Southern California, 34 per cent were made by independents and 55 per cent by the major studios alone or in partnership with other producers. By contrast, of the films made in New York, 41 per cent were produced by independents and 50 per cent by the majors. In Florida, 67 per cent were made by independents, and in Texas, 53 per cent were made by independents. Of the 132 films shot elsewhere in the US, 45 per cent were made by independent and 45 per cent by majors. The major studios clearly have produced the majority of pictures shot in Europe (53 per cent), whereas of the films shot outside of North America and Europe, most (59 per cent) were made by independents.

Thus, feature films made with significant participation by the majors tend to be filmed in Los Angeles and New York; films made principally by independents and filmed outside the US have favored non-European locations, and those filmed within the US have preferred locations away from the traditional centers of Los Angeles and New York. Underlying these differences between the locational behavior of major studios and independent production companies are differences in the economic pressures they face. For example, since low-budget, made-for-television films are a greater proportion of the output of the independent companies, these companies face greater cost constraints. . . .

None of this would have occurred, however, without vertical disintegration, which made it possible for independent production companies to get a foothold in the first place. Vertical disintegration has also affected the locational choices of the major studios, which have shed much of the overhead associated with their facilities and backlots, freeing themselves to carry out production on location. In doing so, they have helped create a new standard of realism in motion pictures that has further strengthened the trend to shooting outside Los Angeles.

Thus, a split locational pattern has emerged in the motion picture industry over the past two decades: establishments and employment have become more concentrated in Los Angeles, while filming has dispersed to other locations.

Explaining the split locational pattern Subcontracting firms have increasingly concentrated in Los Angeles in the past decade because the specialized nature of their services and the constant change in product output requires non-routine, frequent market transactions with other firms. By locating in the center of the motion picture industry, they increase the opportunity to obtain contracts. The transactions ('deals') associated with this process often require face-to-face contact. Production companies and major studios encourage small firms to congregate in Los Angeles in order to ease the managerial coordination associated with the production of a nonstandardized product.

These transactions comprise two basic types; those associated with the search for business and those involving the negotiation of the details of the contracted work. The search for work is a personalized process in the motion picture industry and requires many contacts. Moreover, the relevant decision-making network (e.g., production companies, studio executives) changes rapidly, necessitating constant search to maintain one's network. Negotiation of the details of the production process is similarly unstandardized, for no two motion pictures are alike and it is uncommon for any two production crews to have the same composition. It is much more difficult to carry out these negotiations and the monitoring and information feedback that they require at a distance.

Many subcontracting firms specialize in a particular activity (e.g., recording, special effects) but not in a particular fixed output. They market not only to the motion picture industry but to as many segments of the electronic entertainment

Table 4. Location of production activity

Year	Metro Los Angeles	California	Metro New York	Rest of U.S.	Europe	Rest of World	Total
1960	50[a]	50	5	12.5	22.5	7.5	97.5[b]
	(74)[c]	(74)	(7)	(19)	(33)	(13)	(151)
1965	47.5	47.5	0	2.5	22.5	10.0	82.5
	(93)	(93)	(0)	(7)	(43)	(19)	(190)
1970	22.5	30	10	17.5	32.5	7.5	97.6
	(48)	(62)	(19)	(36)	(67)	(16)	(209)
1975	25	30	5	17.5	27.5	17.5	97.5
	(64)	(81)	(10)	(44)	(67)	(43)	(244)
1980	30	33.5	12.5	22.5	15	15	97.5
	(65)	(74)	(27)	(55)	(33)	(34)	(223)
1984	33	38	13	36.5	6	6	99.5
	(68)	(79)	(28)	(76)	(13)	(13)	(209)

Source: *Daily Variety*
[a]Percent of all production activity in that year at this location.
[b]Total percentages may be less than 100 where some films were not classified. Totals may be greater than the sum of classified films because many films are classified more than once, as they have several locations.
[c]Numbers in parentheses are number of films.

industries as they can. Almost all these firms work in both film and video and produce commercials and industrial and educational films in addition to theatrical or made-for-television movies. Only a few firms concentrate exclusively on television and feature films. Thus, as uncertainty has been transmitted down the subcontracting hierarchy within the motion picture industry, firms on the receiving end have expanded their business opportunities by marketing across industrial boundaries. This increased flow-through offsets some of the uncertainty created by disintegration.

An additional reason for the continued spatial concentration of the industry in Los Angeles is the structure of transactions in the local labor market. Most work in the motion picture industry consists of short-term contracts; individual workers experience considerable variation in and uncertainty about the amount of work they are offered from time to time. Many workers are frequently unemployed, and when work is abundant, they accept overtime to hoard wages against an uncertain future (Storper and Christopherson 1985, Ch. 4). Workers offset the instability of short-term contractual work by remaining close to the largest pool of employment opportunities in the industry. Employers agglomerate to maintain access to a large, highly specialized, and skilled labor force; this allows them to avoid having to retain unneeded labor in downturns.

The mutual interests of workers and employers are met through the process of job search and rehiring. The larger the overall demand for labor in a locality the greater the chance that a worker will be able to secure short-term positions. Of course, most workers in this industry – whether above or below the line – do not participate in an open, undifferentiated labor market. They secure positions through well-established networks of contacts or through institutional means such as the roster system. Even these networks, however, can channel only the work that is available. So, the higher the proportion of the industry's work that is concentrated in the region, the greater the chance the members of search networks will find work (Scott 1981; Spence 1981; Stigler 1961, 1962; Clark 1983). The larger the demand for labor, the more likely it is that an individual worker will secure a job (although aggregate possibilities are determined by the ratio of unemployed to job vacancies in the market as a whole).

In addition, many of the skills demanded in motion picture production are required in other entertainment industries, such as video, television, recording, and theater. To the extent that those industries concentrate in the same place as motion picture production, workers who possess skills useful in more than one industry will have more chances to secure work. At the same time, employers will have access to a larger pool of skilled workers. The emergence of an 'entertainment industrial complex' in Southern California may be making it easier for the different sectors to offset the risks that increasing instability in any one of them imposes on the workforce. . . .

Finally, the strength of external economies within the industry is suggested by the agglomeration of smaller businesses at the intrametropolitan scale: 65 per cent of the industry's establishments (apart from the major studios) are located within the Hollywood district, with the second largest cluster in the San Fernando Valley. This contrasts with the location of sound stage capacity, of which Hollywood has only 13 per cent, Culver City 20 per cent, West Los Angeles 15 per cent, and the San Fernando Valley 50 per cent.

The other side of the locational pattern is the increasing tendency to shoot films outside the Los Angeles region. This new locational flexibility is possible because once the script, crew, equipment, props, and locations have been chosen and financing secured, the level of specialized external transactions declines.

Conclusion

In the motion picture industry, vertical disintegration has followed a phase of oligopolistic competition and a production process that in many ways resembled that of mass production. That this occurred should no longer be surprising; our study confirms the suggestion made by Piore and Sabel (1984) and by Bagnasco (1977) that mass production is not the necessary culmination of sectoral development. Instead, the advantages of internalizing parts of the production process (and all the risk assumption that implies) and of externalizing them (with the loss of control and potential profit opportunities that implies) shift according to changes in the sector's markets, technologies, and labor relations.

Of equal importance is how the historical possibilities for the development of technologies, markets, and labor relations are envisioned. Prior to the introduction of electronically guided, programmable industrial machines in certain industries, for example, minimal optimal scales of production differed from those that exist now and carried with them different pressures for product standardization (Boyer and Coriat 1986). New technologies change scale-efficiency ratios; just as changes in the level of mandated job security alter the potential variability of labor inputs and thus the firm's overhead. The historical debates and the theoretical questions are thus intimately connected to each other. Unexpected change in the former forces questions about the latter. But the reverse is also true: if the border between firm and market is conceived as fluid, then some taken-for-granted models of industrial organization in the postwar period can be questioned, particularly those models that imply that contractual relationships are likely to be important only in the early stages of a product's life.

The vertical disintegration that lies behind flexible specialization creates powerful agglomeration tendencies at the regional level. Flexible specialization itself leads to the recomposition of the industrial complex, through a new form of horizontal integration of production capacity, further strengthening external economies.

Recent evidence from a range of countries supports the observations we have made about the motion picture industry. It indicates that new industrial location patterns are emerging in conjunction with changing forms of production organization. For example, unlike that in the US, the more recently developed Japanese automobile industry is highly agglomerated, as is Japanese television manufacture (Ikeda 1979). In the US, a switch from the 'just-in-case' system of routine materials and information transfer among units of the industrial production system to the 'just-in-time' system of more frequent and smaller-scale transactions appears to require geographical proximity (Estall 1985; Altshuler *et al.* 1984). A tendency toward geographical proximity occurs because transaction costs tend to increase as the scale of transactions is reduced and their frequency increased. In addition, managerial demands for uniformly high product quality increase, requiring frequent monitoring of input providers. Finally, frequent product changes reduce the scale of transactions over time and may require the intensive use of producer services. Thus close coordination between input providers and consumers is needed.

If flexible specialization grows in other manufacturing or producer service industries, then it is likely to be associated with significant changes in existing patterns of industrial location, interurban trade, and interregional growth transmission. There could also be significant impacts on the international pattern of production and trade, calling into question some of the currently fashionable concepts with respect to a 'new international division of labor.' Our findings run contrary to predictions of a steady decentralization of production associated with relatively greater growth in nonmetropolitan areas. Instead, flexible specialization in manufacturing and producer services may help explain the resurgence of metropolitan growth that has taken place in the US in the 1980s.

The geography of technological change in the information economy

M Hepworth, 1986 (*Regional Studies* 20: 407–22)

Introduction

This paper is a vivid illustration of how innovations in information technology will transform urban and regional systems. . . . Spatial systems of information technology may lead to a spatial reorganization of both manufacturing production and office activity. This arises from the technology's potential for overcoming time and distance constraints on production and distribution processes, controlling dispersed operations, and converting material intermediate inputs into electronic form such that they are transmittable over telecommunications lines.

It is shown that computer network innovations have enabled Canadian firms to penetrate new regional markets which, at the international level, are 'unprotected' by tariff barriers. Here, the telecommunications–transportation trade-off is of greater significance than previously assumed, owing to its applicability to both the movement of material production inputs and people.

Finally, analysis of the spatial organization of work in computer-related, printing and publishing, and research and development occupations, suggests that job opportunities in different regions are potentially affected by the 'distance-shrinking' effects of computer networking. Specifically, networking can lead to centralized and decentralized patterns of job-creation, such that today's distributional issues centre on who gets what information and where. . . .

Geographical research

Most geographical literature on the so-called 'information revolution' is still anticipatory (Berry, 1970; Goddard, 1980) or lacks substantive evidence (Richardson and Clapp, 1985). Although 'historical contexts' for situating information technology abound (Rothwell, 1984; Hall, 1985), there is a paucity of empirical work on the current spatial impacts of technical change.

Thus far, innovations in information technology have not led to re-evaluations of traditional theories of regional development, which rest on key propositions about the role of information space and contact systems, the hierarchical structure of diffusion processes, and the locus of corporate control in urban and regional systems (Thorgren, 1970; Tornqvist, 1977; Pred, 1977; Borchert, 1978). Yet, case study and survey research on Canadian multi-locational firms indicates that computer networks enhance head office control over dispersed operations (Hepworth, 1985) and distributed processing technology may induce firms to decentralize management decision-making (Langdale, 1979). Also, network innovations are enlarging information space owing to their 'global reach' and usage by new industries which sell variegated information to firms and governments in the international 'network marketplace' (Helleiner and Cruise Obrien, 1982). Therefore, it is incumbent on researchers to re-evaluate the roles of communications technology and information as a resource and commodity in regional development (Abler and Falk, 1980).

Like telex and the ordinary telephone (Pool, 1977) computer network innovations will transform the space economy. Although some geographers have recognized that telecommunications' role in this transformation is linked to its technological convergence with computers (Goddard et al., 1985), few researchers have proceeded to make computer networks the explicit focus of their empirical work (Bakis, 1985). As a result, the telecommunications–transportation trade-off has remained an elusive and speculative concept (Nilles et al., 1976). By focusing on network innovations fresh insights can be obtained into the key role of telecommunications in regional development. It becomes evident that telecommunications has two vital functions. First, it enables organizations to share centralized computer resources (data, programmes and processing capacity) between dispersed users through a process of information transfer – voice, data, video, and facsimile communications (Tanenbaum, 1981). And second, it enables increases in labour productivity generated by the application of computer technology at one place to be transmitted to and exploited at any other. Thus far, researchers have only considered the potential spatial impacts of voice communications (Goddard, 1975; Goddard and Pye, 1977) and, by considering telecommunications in isolation, have overlooked the importance of the second process. However, it is new non-voice applications of telecommunications technology and the second process (productivity trans-mission) that basically affect the spatial organization of labour and capital at the regional level.

Research on Canadian firms indicates that computer networks tend to be highly centralized structures, although it is now technically feasible to distribute data processing hardware (Hepworth, 1985). In a sample of 117 multi-locational firms from different sectors of the Canadian economy, about 90% of computing capacity was concentrated in metropolitan cities where head offices were located. Users in other cities and regions gained access to central computing resources through telecommunications lines. Assuming that usage of computer power raises labour productivity and given that head offices tend to be concentrated in a few cities, it follows that regional variations in productivity

Table 1. Regional distribution of computer capacity, by sector and province, 1978 and 1983 (%)

		Manufacturing	Oil	Agriculture, forestry, mining	Trans-portation	Finance, insurance, real estate	Trade	Computer services	Federal government	All sectors
British Columbia	1978	5.1	0.1	15.3	2.5	3.4	11.0	1.3	4.0	4.5
	1983	3.8	0.9	20.4	3.8	5.7	13.6	7.9	3.0	6.2
Alberta	1978	3.1	50.8	2.4	0.8	1.9	3.5	0.8	0.8	3.9
	1983	3.5	63.5	9.7	0.5	2.5	1.4	2.0	2.5	7.5
Saskatchewan	1978	0	0	12.1	0	0.2	0.9	8.3	0.4	2.0
	1983	0	0	15.5	0	0.4	0.1	14.4	0.2	3.9
Manitoba	1978	0.4	0	8.3	0.8	9.9	2.3	13.7	2.0	4.8
	1983	0	0.1	4.0	8.4	11.2	2.8	4.6	1.1	3.8
Ontario	1978	78.4	46.8	37.0	3.6	52.9	67.9	57.7	84.7	60.1
	1983	82.4	34.5	31.8	32.7	66.6	69.4	60.4	80.1	64.5
Quebec	1978	12.8	2.3	24.8	92.0	30.4	14.2	15.7	5.5	19.7
	1983	10.1	0.8	9.9	54.5	11.8	11.5	6.3	8.5	12.1
New Brunswick	1978	0	0	0	0.1	0.1	0	0	0.9	0.1
	1983	0.1	0.1	0	0	0	0	0	0.3	0.1
Prince Edward Island	1978	0	0	0	0	0	0	0	0	0
	1983	0	0	0	0	0	0	0	0.1	0
Nova Scotia	1978	0	0.2	0	0	1.2	0.1	0.2	1.1	0.4
	1983	0.1	0.1	0.1	0	1.6	1.2	0.3	3.3	0.8
Newfoundland	1978	0	0	0	0.1	0	0.1	1.2	0.3	0.2
	1983	0	0	0	0	0	0	4.0	0.1	0.7
Total capacity	1978	179.6	18.5	21.3	25.2	83.8	25.2	80.5	57.3	491.5
	1983	1,001.2	296.8	177.3	281.4	621.0	329.3	705.3	432.5	3844.9

Note: Computer capacity is measured by main memory size, in kilobytes.
Source: Sample from *Canadian Computer Census*, 1978 and 1983; Canadian Information Processing Society, Toronto.

will bear little relation to the spatial distribution of capital investment. As such, networking raises serious questions as to how regional economies are modelled, productivity changes are spatially assigned and, by corollary, how and why interregional income and unemployment differentials and factor movements arise. . . .

Technical change in the information economy

. . . Canada's first computer was installed at the University of Toronto, Ontario in 1952. Between June 1966 and December 1983, the national computer population increased from 710 to 16,643, and by now 75% of all computers are used in services, 15% in manufacturing, and the remainder are in primary industries (mainly, oil and gas). The diffusion process has followed the familiar 'S' curve pattern in Canada as a whole and most provinces – that is, a low rate of adoption in the early 1970s, acceleration in the rest of the decade, and slow-down in the early 1980s as the technology matured. . . .

Aggregate data on computer networks used by individual organizations do not exist. Consequently, the regional distribution of computer capacity was inferred through a sample of Canadian multi-locational firms and all federal government agencies. Data on the city location, capacity (as indicated by main memory size) and broad applications of computers used by these organizations is recorded annually by the Canadian Computer Census. It was found that about 64.5% of total computer capacity used in the sample – or the dominant users of network technology – was concentrated in Ontario in 1983 (Table 1). This is consistent with the findings of a similar analysis carried out by the Federal Department of Communications.

There are significant inter-sectoral differences in provincial shares of total computer capacity. Ontario which dominates national corporate and government activity, has consolidated its large share of the technology's usage. However, there has been a significant 'catching-up' effect in Western Canada, where rapid rates of economic and population growth occurred during the 1970s. In contrast, there is little evidence of spatial convergence with respect to the underdeveloped resource economies of the Atlantic Provinces. The most significant trend is the decline in Quebec's share of national computer capacity which occurred in all sectors.

The key factors that explain this pattern of regional concentration are the head office locations and centralized networking strategies of Canadian multi-locational enterprise. About 90% of private corporations concentrated the vast bulk of computing capacity – 80% to 100% on average – at their head offices. Further, this level of concentration did not change significantly between 1978 and 1983, although most firms had decentralized some computer hardware. Thus, Ontario's dominant share of total computing capacity is explained by the structural characteristics of the Canadian urban system, in which Toronto and Ottawa function as national centres of corporate and federal government head office activity respectively. Similarly, Quebec's falling share is attributable to the 'exodus' of corporate head offices from Montreal to Toronto over the last two decades, and Alberta's increasing share reflects Calgary's emergence as a head office centre in the Canadian oil and gas industry (Semple and Green, 1983).

The head office concentration of computer power clearly indicates that networks are essentially instruments of organizational control. A mail survey of Canadian firms confirmed this central role of the technology, which is reflected in the growing importance of strategic network applications (for example, decision-support analysis) relative to routine transactions and administrative applications such as payroll processing and electronic mail.

The same mail survey confirmed that most companies are pursuing centralized networking strategies, although distributed processing architectures are technically feasible. Economies of scale in data processing, which arise from intensive use of mainframe computers, proved to be the salient reason for maintaining centralized networks. Importantly, scale economies are realized by making central computing resources accessible to dispersed users through data communications channels. At present, the overall costs associated with information transfer are of secondary importance and line costs are minimized through the widespread use of concentrators, multiplexers, and other specialized devices. Finally, decentralizing tendencies arising from the adoption of more powerful and cheaper minicomputers at remote locations are not clearly evident. Although some companies have taken this option, networks constructed with minicomputer technology either have specialized applications (for example, Imperial Oil's secondary network for monitoring gas station supplies), or shared public networks and service bureau facilities are used as primary or 'backbone' transmission systems. . . .

Case studies of computer networking

The case studies presented are part of a ten-firm sample which included: IBM Canada, Labatt Breweries, and Massey-Ferguson (manufacturing); Canadian Imperial Bank of Commerce; Imperial Oil Canada (resources); Sears Canada (trade); Air Canada (transportation); and I.P. Sharp, QL Systems, and Infoglobe (computer services). All research material was collected through 1984 from two-stage personal interviews, correspondence, and follow-up telephone conversations. . . .

Case study one: newspaper publishing . . . *The Globe and Mail*, a Toronto-based newspaper, began operating a satellite publishing network in October 1980 following two years of research and development in California, where the *Los Angeles Times* was also experimenting with the technology. The economic motive for networking, whereby laser scanners and earth stations function as specialized computers and terminals, was to increase circulation outside the company's traditional Southern Ontario market, when it faced growing competition from two other regional newspapers, *The Toronto Star* and *The Toronto Sun*. Before the network was established, *The Globe* was printed in Toronto only and about 20,000 copies were distributed outside Ontario by air and road. The basic constraints on regional market expansion were high transportation costs (relative to printing costs and price) and distributing the newspaper in time to meet morning commuter traffic in distant Canadian cities.

The satellite network is integrated into a highly computerized process of newspaper production at *The Globe*'s central printing plant and editorial offices in Toronto (Fig. 1). A new by-product of this process is electronic material for the newspaper's on-line data base subsidiary, Infoglobe. However, the final pre-printing stages are more relevant here, because they involve the creation of an electronic facsimile of each edition of the newspaper which is broadcast to distant printing plants (Fig. 2). Satellite signals are received simultaneously by all plants and converted to photographic negatives from which offset plates for the local presses are developed. This system enables the newspaper to be printed simultaneously across several time zones and distributed from six production sites to different regional markets.

Computer networking has increased *The Globe*'s daily circulation outside Ontario to about 130,000 copies. The average cost of selling the newspaper in other regions has fallen by up to 25% owing to scale economies in satellite

147

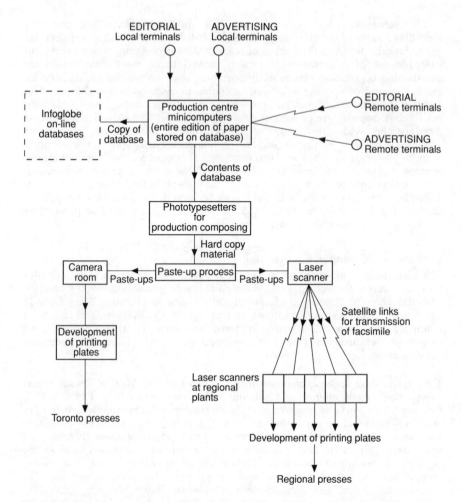

Figure. 1. The organization of newspaper production at *The Globe and Mail*

transmission. At a cost of $700 per hour for transponder capacity, it takes only one and a half hours to broadcast the entire facsimile. Further, with wider circulation, the newspaper has attracted higher-priced national advertising, so that the immediate benefits include lower distribution costs and greater advertising revenue. In the longer term, market entry barriers have been created at the regional level and further expansion will involve outlays only on extensions to the ground segment of the network. The initial investment (excluding research and development expenditures) amounted to about $1.5 million mainly on laser scanners, but three-quarters of these capital costs had been recouped by June 1984.

Additional capital and labour is all informational, however there are important qualitative differences between these resources which are expressed spatially. New investments in traditional capital used by the industry – specifically, printing presses – have been concentrated outside Ontario. The bulk of new electronic capital is centralized in Toronto, where computer technology is

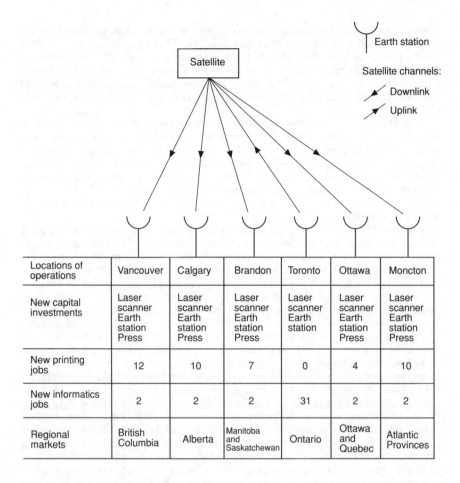

Locations of operations	Vancouver	Calgary	Brandon	Toronto	Ottawa	Moncton
New capital investments	Laser scanner Earth station Press	Laser scanner Earth station Press	Laser scanner Earth station Press	Laser scanner Earth station	Laser scanner Earth station Press	Laser scanner Earth station Press
New printing jobs	12	10	7	0	4	10
New informatics jobs	2	2	2	31	2	2
Regional markets	British Columbia	Alberta	Manitoba and Saskatchewan	Ontario	Ottawa and Quebec	Atlantic Provinces

Figure 2. The topology of *The Globe and Mail* satellite communications network

used in the pre-printing stages of production, to create facsimiles, and to control all operations. Similarly, all forty-five plus new jobs in traditional printing occupations (e.g. pressmen) have been created outside Ontario, but two-thirds of informatics jobs, including all higher-order jobs, are concentrated in Toronto. As such, the relative numbers and types of information employment generated at each location reflect the spatial organization of old and new forms of information capital.

Thus, computer networking has transformed production and distribution processes in Canada's largest newspaper. It has led to radical expansion in market areas, new loci of production, and a geographically dichotomized pattern of job-creation, whose qualitative variations reflect parallel variations in the new information capital deployed.

Case study two: research and development in electronics products Bell Northern Research (BNR), whose head offices and main laboratories are located in Ottawa (Ontario), carries out research and development (R & D) on telecommunications and office automation products for its Montreal-based parent companies, Bell Canada and Northern Telecom. The latter manufactures three product lines which currently take up three-quarters of BNR's manpower resources – telephone switches, circuit-board design, and office automation equipment. R & D is a computer-based, labour-intensive process and 70% of all laboratory work involves creating the software embodied in Northern Telecom's electronic products.

R & D activity is distributed across several research sites in North America and Europe, because marketing of Northern Telecom's products requires that BNR's research teams work directly with the manufacturing company's dispersed customers. Face-to-face contacts are essential for identifying user requirements, ensuring products are correctly installed, and providing post-sales support. These locational imperatives have led BNR to establish new laboratories outside Canada, in the parent company's high-growth American and European markets.

Between 1977 and 1978, a computer network was set up to co-ordinate an R & D process with a unique spatial organization (Fig. 3). A layered approach to product development enables each laboratory to specialize in only part of the work carried out for a specific product line (Fig. 4). For example, base software and hardware for telephone switches are designed and engineered in Ottawa, but user-oriented application software for the same product is developed in Texas and North Carolina. Owing to this inter-regional division of highly-skilled labour, research teams must be kept perfectly informed of changes in product specifications that arise elsewhere. As a result, network traffic consists mainly of very large volumes of R & D information being exchanged between laboratories.

Before the network was established, BNR depended on road and air transport for transferring computer tapes on which product-related information was stored. However, the increased spatial scale and complexity of R & D, arising from BNR's support of Northern Telecom's foreign market expansion, made traditional modes of information transfer inefficient. More trained personnel were required for carrying computer tapes between Ottawa and other cities and supervising customs clearance; more 'paperwork' had to be prepared for customs authorities, placing further demands on technical staff; as new products evolved and international high technology markets grew more competitive, the security risks of transferring R & D information increased accordingly; and, information had to be exchanged in faster time to direct, monitor, and control more dispersed and complex operations. As a 'transport' medium, the current BNR network is a timely and labour-saving system for directly transmitting R & D information between laboratory sites. Transborder data flows are neither detected nor subject to customs scrutiny so that intermediate inputs of Northern Telecom's products are not subject to national tariffs.

In addition to its transport function, the BNR network is used in the direct production of software. Until recently, when some laboratories were allocated computers, all software was developed over telecommunications lines on host systems located at a multisite complex in Ottawa. Some research teams still depend on terminals, which interact with the Ottawa computer systems, and all systems software for R & D computers is developed centrally and distributed over the network. Further, computer systems at distributed sites are maintained through remote diagnostics carried out over telecommunications lines from Ottawa.

Although the spatial scale and complexity of BNR's operations have increased considerably, the bulk of informatics manpower and computer resources used to support R & D activity is still concentrated in Ottawa. There

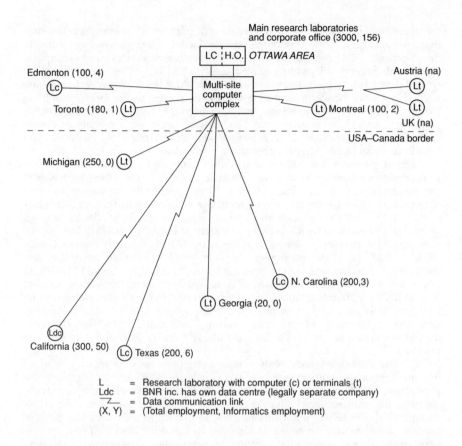

Figure 3. The topology of the BNR computer network

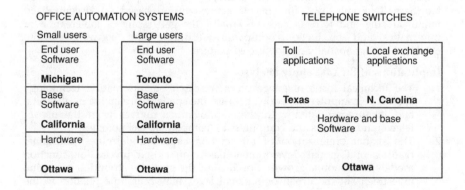

Figure 4. The spatial organization of research and development at Bell Northern Research

are three main reasons for this centralized pattern of informatics resource management. First, scale economies are realized in data processing, software development, and purchasing telecommunications services and computer equipment. Second, all product-related information has to be consolidated on central data bases to co-ordinate research work. And third, physical contacts are needed not only between informatics workers, owing to the problem-solving nature of software development, but also between higher order informatics personnel and line management, to assess user requirements, develop company-wide standards in computing and communications services and contribute to audit and long-range planning functions.

Computer networking has enabled BNR to maintain a centralized approach to informatics resource management because dispersed computer-based R & D can be supported directly from Ottawa. More specifically, scale economies in data processing and software development are realized by distributing the costs of providing these services over all locations. By corollary, the productivity gains of informatics workers themselves are transmitted over the network. Indeed, the number of informatics personnel employed by BNR has hardly changed since 1978, so that the same manpower has been used to provide computer, and communications services to new research teams based in the United States and Europe. The employment outcomes of these spatio-economic processes are that 90% of BNR's 200+ informatics jobs is concentrated in Ottawa; however, led by Northern Telecom's foreign market expansion, 500+ research and development jobs and a 'handful' of 'machine-minding' informatics jobs (for the new distributed systems) have been created outside Canada since the network was installed.

In sum, this case study shows how computer networking can lead to decentralization of production. It has enabled BNR to decentralize the production of intermediate inputs into Northern Telecom's electronic products, because the informational characteristics of these inputs allows for their transmission over telecommunications lines. In this respect, the technical basis of these spatio-economic processes is identical to that identified in *The Globe and Mail* study, where intermediate inputs in newspaper production – editorial, advertising and computing services – were transmitted by satellite as an electronic facsimile. Unlike the *Globe* study, networking did not lead directly to regional expansion in BNR's 'markets' (because R & D services are effectively intra-corporate transactions), but indirectly the company has supported Northern Telecom's own foreign market expansion based on decentralized manufacturing and R & D activity. Finally, the geography of job-creation in informatics and production occupations (printing and software-creation) exhibited similar spatially-dichotomized patterns. . .

Implications of the case study analysis

1. The technical focus of geographical research on information technology should be computer networks, because the spatial reorganization of capital and labour associated with these innovations derives from the use of telecommunications and computers as 'companion' technologies.
2. The spatial organization of information capital undermines traditional theories of regional development based on closed production function models of economic growth. Specifically, the spatial distribution of capital investment, at the urban or regional level, may bear little relation to the geography of productivity change arising from the application of computer technology.
3. The increased significance of the telecommunications–transportation trade-off derives from its equal applicability to the movement of people and

material inputs into production. By focusing on computer networks, it is evident that the expanded role of telecommunications arises from its integration into direct production processes, as well as its traditional function as a distribution medium.

4. Network innovations have created a new form of inter-regional 'trade' in information services, which occurs in an intra-corporate and market context.... As with merchandise trade, the balance of trade in information services will determine how different urban and regional economies are affected in terms of employment, balance of payments (at the international level), local taxation revenues, and industrial development based on local linkage-creation processes. With the growing importance of information services in all sectors, the 'stakes' of developing a comparative advantage in network-based trade increase accordingly.

5. Information-based economies are 'open' economies, in the sense that inter-regional transactions are not yet subject to national tariff policies....

New types of transborder services are emerging as computer network applications are commercially exploited.... The 'network marketplace' is also organized on an international basis ... and existing non-tariff barriers are poor substitutes for systematic regulation of commercial transborder data flow. Although the *Globe and Mail* example is intra-national in scope, identical satellite technology is used by the world's largest newspapers, such as the *Herald Tribune*, for same-day publishing in different countries. In this instance, regional market expansion may be viewed, according to one's ideological orientation, as a new form of 'cultural imperialism' (Galtung, 1982) or as further evidence of the emerging 'global village' (McLuhan, 1964)....

6. While computer networking creates opportunities for decentralizing factory- and office-based activities, policy-oriented analysis should consider these new opportunities in the basic context of corporate structure, behaviour, and strategy. In other case studies, organizational factors explain why programmers at Air Canada's financial offices in Winnipeg (Manitoba) work on computer systems located in Montreal (Quebec), and why all Massey-Ferguson's North American employees depend on computer systems located in the United Kingdom. In the former case, Air Canada's historical links with Winnipeg and the politics of 'high tech' job-location in Canada's crown corporations were key factors in networking strategies and overall informatics resource management. Adverse conditions in North American markets for agricultural machinery has led to corporate restructuring and substantial streamlining of Massey's operations. As a result, computer systems located in Toronto, which formerly supported all Canadian and United States operations, were closed down and satellite communications channels are now used to link North American users to computer systems in England, where Massey has centred its global network.

7. Similarly, while technological developments in distributed processing have led some researchers to suggest that large companies will decentralize management decision-making (Langdale, 1979), Canadian evidence indicates that private companies are currently using hierarchical systems to maintain and support centralized organizational structures. For example, Sears Canada currently operates a Toronto-centred point-of-sale terminal network which supports the retail chain's centralized and mature merchandising strategy. At the local level, terminals provide individual stores with remote access to computer power and information that is essential to routine operations, such as credit authorization and inventory

control; nationally, the terminals provide head office management with 'eyes' at all operations sites through an on-going process by which transactions data is recorded, relayed to Toronto, and stored in central data bases. . . .

8. As the BNR study showed, computer networking can be used to establish a complex spatial division of labour, by serving as a passive medium for transferring information as work material and an active mechanism for controlling and integrating work processes at different locations. It is extremely plausible that network innovations, when used directly to reduce overall labour costs, will lead multi-locational corporations to exploit low-wage markets for information labour, 'de-skill' certain occupations through fragmentation and standardization of work tasks, and re-evaluate the need for face-to-face contacts in selective occupations. For example, access software for Infoglobe's on-line data bases is developed by programmers who work in Toronto computer systems from their homes in Kingston (Ontario), and IBM Canada has conducted a 'work-at-home' experiment with its Toronto-resident word-processing staff. Most importantly, the case studies indicate that 'home' can be in another city or country, as long as telecommunications channels can be used to access working materials – computer power and information. In other words, computer network innovations have drawn attention to the extreme mobility of information at a time when nearly half of Canada's labour force depends on this intangible resource as working material.

9. Finally, the 'information space' in which multilocational organizations operate has been greatly enlarged by computer network innovation. This is brought about directly by physical expansion of private networks, the commercial activities of the 'network marketplace' and government economic-technology policies. For example, BNR's research teams in the United States and Europe have the added responsibilities of monitoring R & D and marketing activities carried out by Northern Telecom's competitors. In Canada, government agencies are using networks to bring together domestic suppliers and buyers, in the belief that 'improvements' to the basic signalling functions of local product markets will lead to new patterns of linkage-creation in the national economy. And, case study research on the 'network marketplace' (I.P. Sharp, QL Systems, Infoglobe) indicates that public data base services are primarily used by multinational enterprise to monitor their global, political, economic and financial environments. . . .

CHAPTER 7

Services in a consumer society

The significance of consumer services

Consumption has traditionally been regarded as the primary basis for service provision. This is why simple economic explanations for service growth relate it closely to rising incomes and consumer expenditure (see Chapter 3). 'Tertiary' activity is still frequently associated with the commercial satisfaction of personal consumer needs, especially through retailing. As we have seen, however, no service function can be so simply characterized. Consumer services certainly do not act alone in providing for needs such as food, clothing, shelter, transportation, entertainment, education or health care. They also require inputs of manufactured goods, of business service functions and domestic and other informal support. Health care, or hairdressing, for example, use proprietary materials and equipment, and recreation and tourism have been revolutionized by manufacturing innovations, ranging from sports equipment to the jumbo jet (Chapter 9). On the other hand, the acquisition and use of manufactured products, such as domestic appliances, require not only formal 'service' support, through design, sales, distribution and repair activities, but also informal 'self-service' effort by consumers themselves, taking a wide variety of forms (Chapter 5, page 227–9).

Consumer services also have differing significance for consumers. Some primarily provide for essential personal needs, including necessities such as basic food, health, education and security, while others supply more discretionary requirements, such as entertainment, sports or tourism. All are supported by communications, transportation and distribution which link consumers to the producers of goods and other services. Such 'infrastructure' or 'circulation' activities have become more important as the scale and complexity of consumption has grown, requiring increasingly varied combinations of goods and services to be available in different places. Technical developments in transport and communications (from the motor vehicle to advanced telecommunications systems) are also associated with many innovations in consumer behaviour (e.g. home shopping, banking and entertainment).

Consumer services also fall on either side of the 'consumption cleavage' between private, market-based provision, and 'collective' or public sector services (Castells, 1977; Pinch, 1985; Warde, 1990). This chapter focuses on the private sector; public sector provision is dealt with in Chapter 8. Nevertheless, urban living is dominated by shared public and private provision in transport and communications, social services, health care, education, administration or recreation. Public provision is especially significant for poorer groups. Indeed, in an era increasingly dominated by private provision, the shift away from public services, through privatization and contracting out, has arguably been as significant for economic and urban restructuring as the decline of manufacturing. Cities are characterized by an intense interaction of different consumption functions. Commercial or political trends which change the balance of service support, therefore, have social consequences extending far beyond the behaviour of individual consumer groups.

This chapter examines the growing influence of commercial consumer markets over modern patterns of social behaviour and service location. These markets are moulded by the tension between the preferences of consumers, seeking to satisfy various wants, and the profit-making strategies of companies providing them with goods and services. In the past, some interpretations, relying especially on neo-classical economics, have emphasized 'consumer sovereignty' in explaining change, while others, especially drawing on Marxist analysis, treated consumption as an extension of producer interests. This distinction has become blurred, however. The independent power of large retailers, not only in relation to consumers, but also to other producers, through investment, advertising and growing market share, has become increasingly obvious (Burton, 1990; Wrigley, 1991, 1987, 1993). Marxist analysis, on the other hand, has increasingly recognized how far trends in late capitalism have become consumer-dominated (Urry, 1990c; see Box 7.1).

The booming modern interest in private consumption has thus attracted a wide range of often incompatible approaches to its analysis. Fine and Leopold convey the resulting confusion:

> Are we the manipulated mannequins of the advertising industry, the sovereignless victims of profit-hungry corporate capital, rational economic men and women trading off one commodity against another according to their relative prices and utilities? Or are we the continuing repositories of custom, culture and family habits, slavishly obeying the imperative to emulate the Joneses? (Fine and Leopold, 1993: 3)

Are consumers victims or in control? Are they guided by rational economic judgements, or by more complex social and cultural motives? Fine and Leopold deplore the disciplinary fragmentation of consumption studies, which leaves such questions unanswered, and

---

Box 7.1: Services and consumption

Urry's work highlights the growing interest in processes of consumption in service studies. He draws on UK analyses of economic restructuring (see Chapter 4; Urry, 1987), but criticizes their emphasis on manufacturing sectors, and the assumptions that the size and shape of service industries simply reflect the local manufacturing base (Urry, 1990c).

Urry elaborates on the role of consumption processes in the shift from Fordist to post-Fordist forms of production (see Chapter 4). He suggests that a new form of economy is developing in which consumption rather than production is dominant (Urry, 1990c, 1990d). New forms of credit permit a growth in consumer incomes, and social change means that there is greater variety and volatility in consumer tastes. Both drive producers to be more market-oriented, and develop more products and services with a shorter shelf-life. This suggests propulsive industries are to be found in consumer-related service sectors, especially leisure, the arts and tourism.

> First, these are highly influential industries. The arts in Britain employ something like 500,000 people and contribute £4 b in overseas earnings (by comparison with 3.8 b from motor vehicles and parts – Myerscough, 1988). Tourism in Britain employs at least 1.5 m people and the industry is worth around £15 b annually. Total leisure spending is worth around £60 billion: five times the total spending on cars (Mills, 1989). Internationally, tourism is growing at 5–6% pa and will be the largest source of employment world-wide by the year 2000. There are absolutely huge flows of tourists internationally.... International tourist receipts increased nearly 50 times between 1950–84, when international tourism had become the second largest item in world trade by value (Urry, 1990a).
>
> Second, such developments are increasingly crucial in determining the character of individual places. Tourism, including arts and leisure, has become central in forming people's experience of place, of the natural built environment, the range of services available locally, the transportation infrastructure, the degree of congestion and the attractiveness that such a place has for inward *manufacturing investment*. ... Nowadays it is almost as though tourism, cultural industries and leisure now constitute *the base* from which all sorts of economic consequences are derived such as new manufacturing employment. (Urry, 1990c: 279–80)

Thus, Urry sees services as an independent source of growth, without strong dependent linkages with manufacturing industry. Some of the trends he identifies could be one component in the development of more autonomous local economies. These could be less dependent on traditional manufacturing exports. More of the spending power of local people could be kept within the region, and households could act as active centres of informal service production (Robertson, 1985; Gorz, 1985; Elkin and Mclaren, 1991).

---

attempt to unify them around a 'production' perspective. They believe that total systems of provision for each commodity need to be examined. These especially embody the varying economic, social and cultural relationships at each of the many stages of production which lie behind the shop window façade, or the breakfast cereal package (Fine, 1993). Another important thread of recent research, however, sees consumption as more than simply the transaction of commodities. Consumers actively interpret the significance of purchases in ways that have little to do with how they are produced and delivered, leading to new fashions and trends in consumer expenditure. Changing consumer patterns are thus created which in a consumer society feed back to lead patterns of production. While a 'systems of provision' approach thus reveals the range of global economic and social power relations which deliver consumer goods and services, it is not the whole story (Glennie and Thrift, 1992). Glennie and Thrift even go so far as to conclude that 'we see consumption shaping production, not the other way round' (Glennie and Thrift, 1993: 604). Thus, modern debate over consumer services still reflects old divisions of perspective.

In this short chapter we can hardly do full justice to this range of views. As we shall see, however, both production and consumption processes are reflected in the changing character and locational patterns of modern centres of consumption. An understanding of the forms taken by investment and employment in consumer services requires insights from both perspectives. We are here primarily concerned with retail location issues, and their implications both for the accessibility of service functions to different consumers, and for the location of service employment. It can be argued, in fact, that the location of consumer spending and its link to the complex spatial distribution of delivery systems and retailing outlets offer a significant integrating focus around which to interpret the disparate perspectives on consumption.

The changing processes of consumption

One reason for the growing interest in consumer services is that consumption has become much more diverse, complex and changeable over the past 25 years (Adams and Hamil, 1991). It is not simply that, with growing wealth, the better-off are able to exercise more choice, as economic theory would suggest. The variety of groups making choices has also grown. Demand has become more differentiated between the rich and poor, men and women, young and old, among various specialist interests, ethnic and other minorities, or urban and rural communities. Social behaviour has also changed, especially in and between the growing variety of household types. Consumer trends have become more unpredictable, and are certainly not simply dominated by particular classes or wealthy groups.

A major consequence of this consumer revolution is that the social identity of individuals now depends on their role as consumers, as well as that as workers (Bocock, 1993). After the 1950s, when mass production and consumption began to dominate economic and social attitudes on both sides of the Atlantic, spending was first based around middle-class home-building and lifestyle, especially in new suburban communities in more prosperous regions. Over time, other groups began to express distinctive consumer preferences through spending and leisure behaviour. These included skilled manual workers, who traditionally belonged to the working class, the growing numbers of well-educated people, women and perhaps especially the young.

Commodities such as clothing, food and drink, vehicles, music and entertainment became increasingly significant indices of social status and change.

For those in employment, the consumption significance of work, and thus of traditional class divisions, has changed with the decline of old manual industries and the rise of a plethora of service occupations. By the 1980s, the lifestyles of the 'new consumers', targeted by market researchers and company sales departments, were identified more by age and life-cycle stage, e.g. the young, singles, married with no children, young and older families or the retired. For many people, work was undertaken primarily to gain the means for expressing individuality through consumption, at home or through entertainment, sport or holidays. While the earnings gap between the rich and the poor may have grown, for both, style of consumption has become the focus of personal achievement, or at least of aspiration. The growing social importance of consumption has also raised the significance of one of women's traditional roles, paradoxically at the same time as their part-time and, to a lesser extent, full-time work opportunities have also increased. But men have become distinctively more conscious consumers too (Savage *et al.*, 1988).

These developments have been accompanied and exploited by a transformation in patterns of consumer service supply, arising from the growing domination of large, sometimes international corporations (Wrigley, 1989). Smaller service firms remain significant only for more specialist needs. Much of the strength and flexibility of these large service organizations arises through their ability to act across various markets. For example, dominant retailing firms, especially in the key food and clothing markets, now bargain powerfully to keep down wholesale prices with the manufacturing businesses whose goods they distribute (McKinnon, 1990). Retail financial institutions, such as banks, also balance their low-risk consumer trade with higher-risk, but more profitable business in commercial and overseas markets. The interests of many service businesses were also fortified in the 1980s by the acquisition and control of substantial land and property assets. This means that consumption trends are linked to (and at times shaped by) global trends and changes in financial markets, even as consumer demands for the exotic have also helped establish world-wide food supply chains (Thrift, 1985).

Control of supply affects the quality, cost and range of choice available to different types of consumers in different places. Suppliers must still anticipate and exploit changing patterns of demand, but they have become increasingly adept at this. The 'new cultural intermediaries' of advertising, the media and sales strategists play on consumers' aspirations for reassurance, excitement, prestige or even environmental responsibility (Gibbs, 1988; Harvey, 1989a). They manipulate not only cost, design and advertising, but also the location and types of environment within which goods and services are made available.

This mutual dependence of trends in demand and supply underlies the flux and instability of modern commercial consumption. Purchases arise from a continuous process of consumer engagement with material goods and services, often indirectly through media imagery. Price is only a broad limiting factor, although it obviously constrains some more than others. Many choices take place between alternatives which cost much the same, differentiated only by presentation, packaging or style. They reflect commercial and other cultural pressures, exerted especially through magazines and newspapers, radio and television, popular music and advertising. Social and personal responses reflect age, gender, personality and past experience, the attitudes of family and friends in different occupational or ethnic groups, and the reputation of particular companies.

As we have already hinted, purchase is also not the end of consumption; in fact, it is only the beginning. Once acquired, consumers 'work' on objects and services, giving them an individual significance (Miller, 1987). They are thus of more than simply utilitarian value; they become part of identity, reflecting personality, self-image, social position, attitude and aspirations. Everyday forms of clothing become items of high fashion, drinking a different type of lager acquires subtle cultural connotations, and esoteric vegetarian dishes become almost as popular as fast food. Particular styles of vehicle, furnishing or vacation may acquire cult status while others appear irredeemably mundane. The attractions even of particular places, such as gentrified inner cities, revived dockland wastes or suburban villages, may also wax and wane in response to popular taste.

This volatility is most obvious for the relatively wealthy, and for such fashion items as clothing, hairstyles, music and entertainment, but even the poor make choices and harbour aspirations from television or magazine images. In fact, cheap or minority forms of clothing, music, or food and drink, reinterpreted through the mass media, pervade the formation of much modern consumer taste. Consumerism is thus dominated by the symbolism of objects and actions. Advertisers, salespersons and shopping developers know this, of course, and devote much of their effort to creating and manipulating images. This cultural and social context of consumption has become increasingly influential in the design and location of shops, retail malls, city centre developments, and holiday and tourist facilities.

The consumer society, however, remains an unequal society. At one end of the spectrum, there has been much discussion of the rise of an affluent and politically significant 'service class' (Savage, et al., 1988; Thrift, 1990a; Featherstone, 1991). This consists of those groups who have the education, skills and experience to occupy well-paid jobs as professionals, administrators, technical experts and managers. Because they remain heterogeneous, however, the career-oriented values of its members are less significant for consumer change, and the physical manifestations of mass consumerism, than their earning power might suggest. They offer a group of 'niche' markets among many others. Even in these terms, they have changed over the past 20 years. Once predominantly in large corporations and public bodies, they now work more in the emerging small firm, self-employed, high-technology and consultancy sectors (Savage *et al.*, 1988). The move towards more mobile, even entrepreneurial, lifestyles has transformed the consumer-based individuality sought by many, whether as an adjunct to work or in domestic and leisure pursuits.

While the influence on wider consumption patterns of the affluent service class may be limited, the plight of the poor and the unemployed is increasingly difficult. The economic restructuring associated with service growth has increased long-term unemployment, especially in areas of industrial and mining decline and the inner cities. The deprivations of unemployment or low pay are made more severe in a society where consumption-based values pervade, and are relentlessly projected through the media. Thus the unemployed and many on low pay are largely excluded from the dominant basis of personal identity and self-expression. Developments in commercial consumer services thus magnify the deprivation of inequality, which cannot be assuaged by 'collective' social services (Little, *et al.*, 1988). It is hardly surprising, therefore, that personal alienation and rising crime accompanies the growth of a consumer society.

Dominant trends in consumer service location

Past studies of the geography of consumer service provision were dominated by ideas derived from Walter Christaller's Central Place Theory (Christaller, 1966; also see Daniels, 1985: 71–103 for a review). This assumed 'consumer sovereignty', with the provision of services basically reflecting patterns of demand. It postulated that services group into a hierarchy of urban centres from which the variety of consumer needs in surrounding hinterlands can be provided for. The supply of 'higher-order' goods, depending on more discretionary spending, needs a larger threshold population. They thus locate in larger centres, drawing on wider hinterlands. Consumers are prepared to travel farther for these than for relatively 'low-order' necessities, offered in smaller, local centres. Consumers' spatial behaviour therefore reflects a 'distance (cost) minimizing' principle, in choosing among the available opportunities.

Central place theory offered a deductive explanation for the variable status and spacing of urban service centres up to the 1930s. As recently as the 1960s and 1970s research still explored the growing complexities of consumer spatial behaviour within this framework. Its application has become progressively undermined, however, by the changing geography of consumer demand, including the effects of growing mobility, and increased consumer differentiation, for example by gender or age, and by the locational strategies of increasingly dominant retail organizations (Bromley and Thomas, 1993a).

Geographical patterns of consumer behaviour since the 1950s have thus been transformed by several well-known trends:

1. The first obvious development has been *the growth of incomes and spending power*. Neither has been geographically evenly distributed. Spending patterns reflect the economic and social composition of regions and cities, and vary with wealth, attitudes to saving and expenditure on regular commitments such as housing mortgages or rents. In the UK, for example, incomes are generally higher in southern regions, but so are basic living costs, especially those associated with high levels of house ownership, and the high cost of property.

2. Regional patterns of spending also reflect *demographic trends*. During the 1960s, consumer changes were significantly influenced by the youth of the post-war 'baby-boom', and this sector has since remained highly influential in consumer sales strategies. Trends from the 1990s, however, are likely to emphasize instead the ageing of the same and subsequent generations, and their enhanced spending power. Equally significant social changes have transformed households into many forms other than the nuclear family. As we have seen, this has created a much greater diversity of lifestyles and patterns of aspiration. One consequence has been a loosening of individual ties to particular places, and the greater tendency to move to new locations, both within current regions of residence, and farther afield. Another has been problems of access to consumer services for groups such as single parents, or young mothers with children, especially where they have no car (Bromley and Thomas, 1993a).

3. A universal geographical consequence of these changing demographic patterns and social attitudes has been *counter-urbanization*, or *population decentralization*, from the major cities. Extending beyond their former suburban development, this trend has generally been led by more affluent groups, seeking smaller town or rural living environments. In many cases, the shift was initially based on commuting back to the city for work, and to acquire many consumer and other services. Once established, however, 'exurban' developments have increasingly attracted their own service and employment opportunities. A critical part of this process, of course, has been the switch from dependence on public transport to almost universal household car ownership. This has placed the population in increasingly congested older urban areas, built during earlier public transport-based eras, at a disadvantage. The infrastructure of high-volume roads around cities now provides a framework upon which new automobile-based centres of consumer spending may be arranged. These shifts were initiated in North America even before the Second World War, but have become predominant in Europe since the 1960s, constrained only by planning, environmental and conservation concerns.

4. Another important facet of modern consumer demand reflects *changing patterns of employment*, many of course related to the growth of services themselves. With high proportions of married women working, often part time, the convenience of car-based,

super-store shopping, unconstrained by traditional retailing hours, has provided a powerful competitive lever for consumer service businesses. The attractions of traditional high-street shopping have waned, as out-of-town centres offer a variety of outlets in a secure, often weather-proofed environment (Davies and Champion, 1983; Dawson, 1988; Carvey, 1988; NEDC, 1988; Schiller, 1988; Robertson, 1989; Jones, 1989). In some cases, as we shall also see, the 'recreational' aspects of shopping are also being served through the provision of more diverse environments, designed to exploit these aspects of consumption, often projected in the home through the media.

5.   The *location strategies of retail organizations* over the past 20 years have been designed to exploit these radical changes in patterns of consumption to their best advantage. Established large retailers, especially traditional city centre multiple stores, have had to adapt the location and quality of their services. Where successful, they have provided the core stores in new out-of-town locations, often closing old downtown sites. Generally, the most successful firms have specialized in particular types of goods, whether food, clothing, household goods, furniture, carpets, DIY gadgetry, automobiles or even toys. Initially they emphasized cost competition, based on the volume of turnover they could command. Their strategies of investment and growing market control have been increasingly based on manipulating the whole chain of activities which deliver goods and services to consumers, including supply and market intelligence, storage, distribution, advertising and retailing.

As a few major firms have come to dominate regional or even national markets, competition for market share has shifted to strategies of pre-emptive site development, and a greater emphasis on the quality and range of goods. Further technological advances have enabled the distribution process to be even more tightly controlled. Electronic point of sale (EPOS) registration now allows sales to be linked in great detail to stock control and monitoring. This in turn allows labour costs to be attuned more closely to income (Chapter 5). Cashless, plastic card-based shopping is also becoming common, and the spread of cable and satellite-based information systems will allow home-based shopping to become more widespread. Large organizations have thus been responsible for major shifts in the form, including the geography of consumer service provision. In the US, for example, the largest company, Walmart, employed 365 000 workers in 1992. The leading supermarket chain in the UK, Sainsbury, employed 85 000, while Marks and Spencer, clothing and food, employed 65 000 (Bromley and Thomas, 1993b).

6.   These market developments have been to some extent encouraged and moulded by the *different regulatory framework of each country* (Blomley, 1987). In the UK, these have been dominated by planning limitations on the development of large greenfield sites, and there

have been only limited objections to national market domination. Even such constraints were generally relaxed during in the 1980s, partly explaining the pattern of rapid expansion. In the US, commercial anti-trust regulation is generally stricter, at least within the major regional markets. Continental Europe has seen the development of more aggressive cut-price chains than in the UK, but the generally higher density of urban living has also sustained town centres better. Patterns of retailing have also been increasingly influenced by wider social and political attitudes, for example, leading to the scrutiny of the origins of produce, and the development of environmental awareness.

A new geography of retailing has thus emerged in Europe since the 1960s, largely following trends in the major regions of the US, but based on national competition between dominant companies (Brown 1987; Ducatel and Blomley 1990). Emphasis has decisively shifted to 'out-of-town' retail locations, oriented to motorways and automobile access. In Europe, however, many older urban centres have retained a significant share of business by developing car parking, pedestrianization and specialization of functions. The last trend is sometimes associated with tourist and other cultural attractions. Traditional retail centres have thus had to offer the convenience of the new centres, with redevelopment in many of the larger cities taking the form of shopping malls, associated with other types of property development, including offices and tourist facilities.

In the UK, the old hierarchy of consumer service supply has thus been supplemented, and partly displaced, by various alternative forms of development (Bromley and Thomas, 1993b). Individual food-based superstores of up to 4650 sq. m (50 000 sq. ft), and occasionally even the larger hypermarkets favoured in continental Europe, are now scattered on accessible sites throughout and around the outskirts of urban centres. These are supplemented by a wide variety of retail warehouse outlets, of over 930 sq. m (10 000 sq. ft) selling bulky goods, including DIY, electrical goods, furniture and carpets. Both types of outlet are sometimes grouped into retail warehouse parks, with three or more units. More consciously planned retail developments include subregional shopping centres of up to 37 200 sq. m (400 000 sq. ft), combining a superstore, another large retailer and a variety of smaller shopping units.

The ultimate manifestation of this form of development is found in regional shopping centres, of over 37 200 sq. m, which have been developed as fully integrated shopping malls, including a range of multiple stores and other outlets, and incorporating other functions, including leisure activities, and office and other commercial developments. These centres challenge traditional urban central business districts in both their size and range of facilities. Following the pioneering example at Edmonton in Alberta, Canada, these 'mega-mall' developments include the MetroCentre in North East England, Merry Hill in the West Midlands, Meadowhall near Sheffield in South Yorkshire, and the Lakeside Centre at Thurrock, east of London. Significantly, like some of the warehouse outlets, these have been sited on derelict mining or industrial sites, encouraged by regional economic revival strategies, including 'Enterprise Zone' investment incentives. Centres of consumption, rather than production, have thus increasingly been employed to provide a dynamic basis for regional economic restructuring. There is, however, little evidence that extra jobs were created by these developments, once the diversion of activity from older centres nearby is accounted for (Lowe, 1991b).

UK food retailing

The UK food retailing sector is a good example of the dominance of large retail organizations and their influence on location (Wrigley 1987, 1991, 1993). There are several reasons why concentration trends were particularly evident in UK food retailing during the 1980s:

1. The adoption of computer-based innovations in supermarket technology allowed a hugely increased volume of trade at much lower unit costs.
2. Extended opening hours and the flexible employment of workers maximized customer convenience, while controlling labour costs in relation to daily, weekly and seasonal fluctuations of demand.
3. Supply networks were transformed through the strategic location of intermediate warehouses, operated and monitored by computer methods, linked to superstores by regular road delivery. Much of this distribution was competitively contracted out to specialist hauliers, and sometimes manufacturers were also expected to deliver regularly to major stores.
4. The introduction of cut-price, 'own-brand' versions of many basic items also placed competitive pressure on manufacturers. Retailers thus increasingly became key coordinators of manufacturing as well as distribution functions. In some cases, major retailers even exert supervisory control over the manufacture of the goods they sell (Ducatel and Blomley, 1990: 218–25).
5. A further influence on change during the 1980s, however, was government policy. This relaxed restrictions on the development of large sites in prime suburban locations and on the outskirts of cities and towns. At the same time, companies were allowed to attain dominant shares of the national market without being significantly challenged by monopoly, or anti-trust legislation. In the **reading** (pages 171–7) Wrigley summarizes the supportive views of both the Monopolies and Mergers Commission and Office of Fair Trading on this. The official view was that the level of competition was adequate, and the quality of the new breed of stores justified their growing domination. The whole process was fuelled by the 1980s property boom, which at one stage enhanced the asset value of the companies even as investment took place in new stores, allowing

them to acquire new capital through 'leaseback' arrangements with banks and property companies.

Wrigley focuses on the strategies behind this growing dominance of three companies in UK food retailing; emphasizing the role of firms in shaping spatial patterns of retail development. He especially emphasizes the significance of access to large-scale capital resources in the transformation of retail geography. This, of course, emphasizes the links between finance, property and retail development, and requires that retailers maintain the confidence of lenders, such as pension funds, banks and other investors through the stock market. Wrigley speculates on the future of this process, in view of the very high profit expectations set by the 1980s experience. He argues that excess profits were made by the major companies during the late 1980s, based especially on their increasing power to dictate the prices from suppliers. They also moved 'up-market' into fresh foods, delicatessen, bakery and higher value-added processed foods, where profits are higher. This strategy may nevertheless leave them vulnerable to competition both from European rivals, seeking their share of excess profits, and from cheaper, down-market, low-cost 'warehouse' outlets.

Even in the recession of the early 1990s, the companies were planning many new sites, with new configurations and ranges of products, and moves into regions where they are still under-represented. Further cost cutting was planned, especially through technical changes in scanning, monitoring and stock control, reducing storage requirements. While the permissive regulatory environment of the 1980s was changing, with stricter planning control on sites nearer smaller towns, scope for expansion remained, if only into overseas markets. During the 1980s competition was based on store location, quality and a growing range of goods. In many areas, Wrigley points out that investment strategies created local market domination, rather than price competition. Continuing developments depended on access to capital, and thus on investor's expectations of profit growth. By the mid-1990s, expansion plans were being further reviewed, in the light of reducing returns to further investment, especially as low-cost retail warehouse companies entered the fray.

Consumer culture and the built environment

We suggested earlier that the growing dominance of major service suppliers may to some degree be balanced by the growing complexity of consumption itself and, as some would argue, the growing power of the modern consumer. Modern consumerism is driven by the speed and geographical range of communications, drawing images from a complex urban-dominated culture, linked to the global media. This has hugely increased the range of both commodities and images of consumption, and accelerated the pace of their transformation and re-interpretation in a plethora of social contexts (Tomlinson, 1990). If this is also accompanied by growing inequality in incomes, however, especially as former 'luxuries' such as washing machines or motor cars become near-necessities, poorer groups are increasingly disadvantaged.

The most distinct geographical outcomes of this supposedly 'post-modern' condition in the advanced economies have been the new superstores and shopping malls, and the 'downtown megastructures' of US and, increasingly, of European cities. These are designed to associate commercial appeal with diverse cultural myths, encouraging consumers to treat them as centres of wider exchange and discourse, and as sources of information and entertainment. The goal, especially in the US, is to create a sense of place, often by artfully designed association with traditional models of social activity, perhaps small-town or rural, medieval or western, 1890s or 1920s (Goss, 1992; Harvey, 1989a).

The **reading** from Goss (pages 177–83) is taken from a longer chapter on retail developments in the US. This includes a commentary on the symbolism of the built environment, linking the current phases of capitalist evolution, especially the transition from Fordist to post-Fordist modes of production, to modernist and post-modern architectural trends (this perspective has much in common with that of Harvey, described in Chapter 4, though it attributes greater significance to trends in consumption). The selection from the chapter focuses on the consumption patterns which parallel these developments. The evolution of the US retail built environment reflects both a symbolic and material transformation in the processes of consumption. In the US, Kowinski's 'Malling of America' has replaced older urban service centres more completely than has so far been the case in Europe. This has not simply been for retail functions, but also for most other types of private and community service. The shopping mall offers a weather-proofed, automobile-oriented, secure environment, within which consumer preferences of many types can be reflected and exploited. Thus, in the US the culture of modern consumption has more completely engulfed the traditional geography of service provision than elsewhere.

In North America, Goss sees suburban malls, and the integrated, high-rise office, entertainment and shopping complexes which have become the downtown alternative to them, as temples to consumption, serving the new middle class of discerning consumers. Shopping has become an aesthetic, and often historically 'themed' experience. He suggests that consumer spectacle and consumption, rather than production, have become the driving force of social life. The interaction that we have described between dominant supplier interests, in retailing and transportation, and the diversity of consumer preferences and perceptions, creates powerful cumulative forces for economic growth in favoured areas (Box 7.1). These processes, however, disadvantage poorer groups, and poorer areas within regions or cities. This has important implications for local economic development strategies (Chapter 9). Lavish consumer display reinforces stark lifestyle contrasts, for example, between inner and outer urban areas, or between city centres dominated by carefully controlled office, retailing and tourist complexes, and nearby areas of inner city deprivation. New developments, as we have suggested in the UK, drain away support for local services in nearby centres, adversely affecting the poorer, older and less mobile.

Like Fine and Leopold earlier, Goss also points out that the emphasis of modern consumerism disguises the social nature and origins of goods, in the labour of others, wherever it is employed in the world, and under whatever conditions. Even locally, the 'theming' of outlets obscures the conditions of much consumer service labour. Service workers must increasingly conform to the highly prescribed, almost 'theatrical' requirements of working in fast food restaurants, or tourist and entertainment facilities. More routinely, even retail and banking outlets, for example, increasingly depend on automated and regulated systems of delivery, but staff must still primarily appear to offer personal service, in a manner acceptable to target groups of customers. This favours certain types of worker over others, as we saw in the Crang and Martin reading in Chapter 5. In spite of this, many consumer service jobs are becoming more routine, less individual, more insecure, less well paid, more often part time, and less satisfying for many workers.

Popular culture, of course, may be the source of much innovation and energy, but its commercialization is also associated with the polarization of access to wealth, especially through its dependence on low-cost labour. Employment in many private consumer and public services has tended to stagnate or decline in recent years, in spite of apparently growing demand. Many functions, especially transportation and retailing, have been automated to reduce costs and serve large-scale needs. The share of full-time workers has also been reduced, as we saw in Chapter 2, in favour of part-time or temporary employment, more of it for women than for men. Better-off consumers, the target of modern consumer services, are thus free to choose, and even to innovate in their behaviour in response to the choices presented to them. Others have poorer access, or are subordinated by their dependence on poorly paid service work (Lowe, 1991a,b). Dominant consumerism thus emphasizes status and individualism over collectivism, and private affluence may be accompanied by public squalor, as J. K. Galbraith highlighted 30 years ago (Chapter 8).

The symbols of modern life are consumption, leisure and lifestyle. The reality for many, of course, is very different. Nevertheless, TV watching, fast food, alcohol consumption, mass sports, motoring and theme parks are almost universal. Even history is decreasingly sought in museums, art galleries, or real towns and villages, but in heritage experiences. Tomlinson (1990: 21) comments that 'Fewer and fewer people in modern capitalism work at making things. More and more people work to make impressions', echoing Ewen's characterization of modern entrepreneurs as 'captains of consciousness'. Marketing flatters consumers; objects are made significant through packaging and advertising. Plazas, malls, market-places and the media are dedicated to the seduction of the consumer.

Some implications

The analysis of consumption has evidently always formed an important part of the study of service location. In an advanced consumer society, however, a much wider range of economic, social and cultural influences needs to be recognized than in traditional service location analysis. In some cities and regions growth may even be consumer-based. As Urry argues, 'propulsive enterprises in the leading sectors [of the city] are to be found not in extractive or manufacturing industries but in consumer-related services' (Urry, 1990c: 279; Box 7.1). Cities are centres for the consumption of leisure, recreation, tourism, sports and the arts, as well as being retailing centres with a regional or even larger market. Though many such consumer functions rely heavily on local income and expenditure, during the 1980s they grew at times without many obvious connections with the local economic base. Unlike tradable sectors they also benefited from low levels of competition through imports from elsewhere. Indeed, the processes of economic development appeared to work in the opposite direction to that envisaged by traditional economic base theory, with the image created of the city by recreation and cultural activities attracting outside investment into the local economy.

The 1980s were a particularly favourable period for the expansion of consumer services, fuelled by the growth of consumer expenditure as a proportion of national income, and a widespread increase in personal indebtedness. The late 1990s are unlikely to witness such a consumer boom, as the problems created by indebtedness in the 1980s (e.g. in property) take time to unravel. In such a climate the limits to the capacity of even the very largest retail schemes for employment creation will become more apparent, especially with employment losses elsewhere in local economies (Lowe, 1991b). Nevertheless, while the local economic bases of manufacturing, finance and business services are important influences for any urban economy, the evidence of the 1980s shows there is considerable scope for independently supported consumer service growth.

The chapter also shows the need to integrate economic, social and cultural processes in the analysis of local economies. For example, the literature on economic restructuring, which has been so influential in Britain (see Chapter 4), analyses localities in terms of their place in the spatial division of labour, determined by the sectors and organizations historically present in the area. Different patterns of restructuring, stemming from different imperatives of capitalist accumulation, are laid down to produce the unique local response to broader trends. We are now in a position to see that such analysis must accommodate more than economic change to understand local dynamics. This obviously includes an analysis of race and gender, but must also go further, to analyse cultural trends and the social milieu associated with consumption. This conclusion parallels the inclusion of informal activities in work in Chapter 5. Just as the analysis of service work should extend outside the formal economy (Chapter 5), so consumption has a wider significance than the distribution and sale of commodities.

Further discussion and study

Guided readings

Bromley R D F, Thomas C J (eds) 1993 *Retail change: contemporary issues.* UCL Press, London

Carvey R 1988 American downtowns: past and present attempts at revitalisation. *Built Environment* **14**: 46–70

Dawson J 1988 Futures for the high street. *Geographical Journal* **154**: 13–16

Glennie P D, Thrift N J 1992 Modernity, urbanism and modern consumption. *Environment and Planning A, Society and Space* **10**: 423–42

Little J, Peake L, Richardson P (eds) 1988 *Women in cities.* Macmillan, London

Robertson K A 1983 Downtown retail activity in large American cities. *Geographical Review* **73**: 314–23

Wrigley N 1993 Abuses of market power? Further reflections on UK food retailing and the regulatory state. *Environment and Planning A* **25**: 1545–52

Activity

1. Examine a modern shopping mall or centre and compare it with a smaller more traditional shopping centre either in a suburb or small town:

    (a) What types of people use it (income, age, ethnic origin, gender, locals, tourists) and what effect does this have on the types of goods sold there?

    (b) How many shops are there; what proportion are owned by large retail chains? How many are locally owned?

    (c) How much competition is there? Does more than one store sell particular goods? What scope is there for comparative shopping? Where is the nearest shopping centre selling the same goods?

    (d) What proportion of people are actually shopping (e.g. how many are carrying purchased goods), and how many are just looking, relaxing or accompanying purchasers?

    (e) What sort of traffic access is there, for those with and without cars? Where does public transport link to, and with what frequencies?

    (f) What provision is made for children, old people, teenagers, the disabled?

    (g) Is there evidence of security surveillance and control?

    (h) Does the centre have a distinctive image or atmosphere? How far is this consciously designed? What is this, who does it appeal to, and why? How is it advertised?

(i) Are non-commercial, community services also provided?

Write a report advising a major retailer whether they should set up in the centre. What are the advantages and disadvantages of the centre? What sales strategy should the retailer pursue?

Consider

1. Why did major retailing firms attract such huge amounts of new investment during the 1980s?
2. Summarize the processes which Goss suggests explain the emergence of the 'post-modern consumer spectacle', replacing the traditional shopping centres. How universal is this trend (think of your local centre)?
3. What is the relationship between Goss's analysis of consumption and studies of 'post-Fordism' discussed in Chapter 4, or 'flexible specialization' in Chapter 6?
4. What factors should be taken into account in measuring the impact of a major new shopping development on employment in surrounding areas?
5. In the light of technological change, what prospects are there for the growth of employment in consumer services?

Discuss

1. Major shopping schemes are now often regarded as a valid form of urban and regional economic development, attracting investment and creating jobs where none existed before. In what circumstances might this be true, whether with or without other attached business or recreational activities?
2. 'Modern consumer practices are designed primarily to disguise the origins of goods and services, and the nature of conditions under which they are produced.' Is this too harsh a judgement?
3. Is the expansion of out-of-town retailing leading to the death of the city centre?
4. Recent changes in retail location have reduced the accessibility of consumer services for less well-off groups. Discuss.

Readings

Is the 'golden age' of British grocery retailing at a watershed?

N Wrigley, 1991 (*Environment and Planning A* 23: 1537–44)

Raising capital in a services-led recession

... It is often said that the current, alarmingly deep, recession in Britain has very different characteristics from the recession of 1979–82. It is a services-led and property-led recession whose urban and regional pattern of impact is almost the mirror image of that of 1979–82. As such, it is the boom areas and sectors of Mrs Thatcher's Britain, the south and east, the financial sector, property, retailing, and so on, which have seen the greatest rises in unemployment and falls in exchange values. And it is against this background which the events that have prompted this commentary must be set.

On 18 June 1991, J Sainsbury plc, Britain's largest grocery retailer asked its shareholders, via the London Stock Market, to subscribe an extra £489.4 million to fund a store-expansion programme. In itself this event was unusual – being the first ever public rights issue in the firm's 122-year history. But it was given extra significance by the fact that Sainsbury's two main rivals, Tesco plc and the Argyll Group plc (the operators of Safeway) had both launched similar cash calls earlier in the year – Tesco calling for £572 million in January and Argyll for £387 million in May. At first sight, raising a total of £1.4 billion at the very bottom of a services-led recession might seem a formidable task, particularly when in the Sainsbury case the Sainsbury family were not prepared to invest a single penny of new money in the rights issue and were, in fact, net sellers. It is a testimony, therefore, to the extraordinary performance of Britain's leading grocery corporations during the last decade (for example, in the case of Sainsbury: net profit rises of more than 20% per annum sustained throughout the 1980s; annual net profit now exceeding £500 million; shareholder-dividend increases of more than 20% per annum for the past twelve successive years, and so on) that none of these cash calls has experienced the slightest difficulty.

Moreover, it should be remembered that the rights issues of 1991 are only one way in which the leading grocery corporations have been active during the current recession in raising capital.... In the case of Sainsbury, two other techniques have been used to supplement retained profits. First, stores have been sold to property companies and then leased back – Sainsbury has raised over £300 million in this way during the two-year period 1989–91, leaving 60% of its stores owned outright and 40% rented. Second, capital has been raised on the bond markets.

Grounding the capital

Why then should Britain's largest grocery retailing corporations be such voracious consumers of capital in such uncertain times? With the exception of Asda (Associated Dairies Group plc), more of which below, it is *not* because of any financial troubles. Neither is it to fund recessionary 'bottom fishing' expeditions. It is simply, as stated by the corporations themselves, to finance organic growth through new-store development programmes; programmes which have become ever larger over the past decade. In the case of Sainsbury, for example, annual capital expenditure (primarily on new-store development) increased no less than tenfold during the 1980s to over £500 million per annum by 1989; an increase five times greater than that in the retail prices index. In 1991/92 this capital expenditure will exceed £800 million per annum, and the capital expenditure programme of Tesco will have increased even more rapidly to over £1 billion per annum.

The reasons for this centrality of the new-store expansion programmes are the ones I have drawn attention to in my previous commentaries. Quite simply, the largest grocery retailers generate no less than *two thirds* of their annual increases in sales from the new stores which they have opened in the previous twelve months, and those new stores typically operate at higher net profit margins. Thus the new-store expansion programmes have become vitally important in maintaining the substantial annual increases in turnover and profit which the capital markets have come to expect.

In a very real sense, the critical arena of competition between the major grocery retailers in the United Kingdom has become the new-store development process, and has long since replaced price competition in this role (the 1980s and early 1990s have truly been an era of 'store wars' rather than 'price wars'). Given the rationing of sites for large-scale development by planning constraints, particularly in strategic locations, competition between the major grocery retailers in Britain is at its most intense during the struggle for the most attractive development sites – ideally, sites which 'offer high catchment-area expenditure but limited competition from stores of similar size and vintage' (Moire, 1990, page 108). Once a store has been built and is trading (having cost on average about £25 million in 1991 prices to develop the typical new Tesco/Sainsbury/Safeway store of around 35 000 ft^2) the intensity of competition declines, and, in practice, competition is almost nonexistent once a shopper chooses (is 'captured' within) a store. As Moir (1990, page 112) has perceptively argued, in the current oligopolistic trading conditions in British grocery retailing, 'profit maximising occurs if larger retailers contest for store sites but cooperate once those sites have been secured. Ensconced and dominant in their own market, new large stores enjoy a degree of protection from all.'

So, as the arena of real competition between the majority grocery retailers has become ever more firmly located in the new-store development process and in strategic capital investment, differential access to capital to fund development programmes which now significantly outstrip what can be accomplished from retained earnings has become an increasingly vital issue. Arguably, it is the differential ability to raise and ground capital amongst the major grocery retailers which offers one of the keys to understanding the future trajectory of this sector of the British service economy.

A transformation in the concentration of capital

Capital expenditure programmes of the scale envisaged by Sainsbury, Tesco, and Argyll for the early 1990s, must clearly impact upon market shares. The story of the 1980s, as outlined in my previous commentaries, was the rise of the 'big five' group of grocery retailers: Sainsbury, Tesco, Argyll, Asda, and Gateway (previously Dee Corporation). Although measures of market share are notoriously confused and confusing in this field, the rise of this 'big five' can be represented to a reasonable degree by the figures shown in Table 1. By the end of the decade the combined market share of these firms had more than doubled to approximately 58% of total British grocery sales. However, it was widely believed in the British retailing industry that there was not room for five major players in the grocery market.

Subsequent events have borne out this view. In late 1989 Gateway was acquired by the Isosceles Consortium. To service its debts, Isosceles disposed of assets (including sixty-one superstores to Asda for £705 million and forty-two smaller stores in Tyneside and Humberside to Kwik Save for £26.5 million) and restructured the remaining parts of the firm into a smaller operation. Asda, seeking to increase its market share via the purchase of Gateway superstores, overextended itself. As a consequence, and exacerbated by the poor returns of its MFI furniture division as the recession began to bite, its performance began to suffer. Net profits began to decline (approximately £165 million is expected in 1991 against £247 million in 1989), the capital expenditure programme went into reverse (only £200 million in 1991/92 compared with Tesco's £1 billion), the MFI division was shed, its share price fell and, in June 1991, its chief executive left the company. Asda is now struggling under a mountain of debt. It urgently needs a refinancing package if it is not to be forced to sell off some of its choicest sites to its chief rivals.

The picture for the early 1990s is, therefore, the likely replacement of the 'big

Table 1. Top five British grocery corporations: estimated shares (%) of total British grocery sales.

	1982[a]	1984[b]	1988/89[c]
J Sainsbury plc	9.5	11.6	14.5[d]
Tesco plc	8.7	11.9	14.8
Dee Corporation/Gateway plc	na	7.3	11.4[f]
Argyll Group plc	(3.8)[e]	5.1	9.7
Asda (Associated Dairies Group plc)	4.6	7.2	7.9

[a]Source: AGB figures quoted in Davies et al (1985, page 9).
[b]Source: Verdict Market Research.
[c]Source: *Retail Business Quarterly Trade Reviews* number 12, December 1989, page 12. Figures relate to year to March 1989.
[d]Does not include grocery sales of Savacentre, the Sainsbury subsidiary. If included, rankings of Tesco and Sainsbury are reversed.
[e]Crude estimate: Argyll acquired Allied Suppliers in 1982. Allied's market share on acquisition was approximately 3%.
[f]Prior to acquisition of Gateway by Isosceles. Since acquisition and disposal of assets, market share has fallen considerably.

five' by the 'big three'. The rights issues of 1991 merely confirm this. They can be interpreted as part of a predatory game which is being played out with Asda and Gateway cast in the role of the prey. Certainly, several commentators have suggested that the timing of the Sainsbury rights issue was as much to do with denying Asda access to its vitally needed refinancing as to meet Sainsbury's avowed new-store expansion objectives. In that sense, it merely confirms the views expressed above. The locus of real competition between the major grocery retailers is passing ever further up the chain of strategic capital investment. In the early 1990s it is being fought out as much in terms of differential access to capital as it was in the 1980s in terms of the struggle for the most attractive development sites, or in the late 1970s and early 1980s in terms of access to price advantage via increasing control over suppliers.

Who pays? – the issue of 'excess' profits

The consistently rising net profit margins of the leading grocery retailers during the 1980s (see Table 2) increasingly roused the suspicion that 'excess' profits are being made in British grocery retailing and that British consumers are paying ever more dearly for the growing concentration of capital within the industry.

This is a highly contentious issue. In the late 1970s and early 1980s, the

Table 2. Trends in the net profit margins (%) of Britain's top grocery retailers during the 1980s (source: company reports; figures based upon turnover exclusive of VAT).

Retailer[a]	1983	1984	1985	1986	1987	1988	1989
J Sainsbury plc	4.6	5.2	5.3	5.7	6.5	6.9	7.1
Tesco plc	2.7	2.8	3.0	3.9	5.0	5.7	6.1

[a]Sainsbury and Tesco traditionally report their net profit margins on a different basis: Tesco on VAT-exclusive sales and Sainsbury on VAT-inclusive sales. In this table the figures are standardised to a VAT-exclusive basis, and profits are before employee profit sharing and taxation.

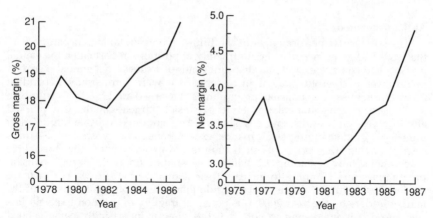

Figure 1. Larger grocery retailers' gross and net profit margins 1975–87 (updated from Moir, 1990; primary sources: *Business Monitor* SDO and SDA 25, Office of Fair Trading, and Institute of Grocery Distribution).

Monopolies and Mergers Commission (1981) and the Office of Fair Trading (1985) examined the impact of increasing retailer concentration on the nature of competition in the British retailing industry, in particular the increasingly oligopsonistic buying power of the major grocery retailers. In both reports it was concluded that the increased buying power was not harmful to the public and that it had been part of a parcel of developments which had been beneficial to competition and to the consumer. Moreover, the leading grocery retailers adamantly maintain that the sector remains intensely competitive. They argue that increased margins merely reflect added value and changed commodity mix, that is that they are successfully selling greater amounts of traditional high-margin items such as fresh foods. Furthermore, they argue that, in comparison with the USA, for example, higher margins are necessary in Britain to repay the higher investment required on expensive British land. Nevertheless, in international terms, the net profit margins being obtained by Sainsbury and Tesco are quite extraordinary (see Wrigley, 1989). And they certainly lie at the root of the German retailers' desire to penetrate the British market; as illustrated by the recent entry of Aldi into the limited-line discounter segment of the UK grocery market previously occupied by Kwik Save and the Lo-Cost division of Argyll. . . .

Figure 1 shows the pattern of gross margins and net margins amongst larger British grocery retailers from the mid-1970s to the late 1980s. What is clear is that, when the Office of Fair Trading (1985) conducted its investigation of competition in British retailing, drawing on information for the period 1975 to 1983, there was no evidence of marked and sustained increases in net profit margins nor evidence of the emergence of 'excess' profits. But, clearly, the period from 1982 to the end of the decade shows evidence of very different conditions and the strong possibility of the emergence of 'excess' profits. Moreover, Figure 1 indicates the possibility that under a different political regime in the United Kingdom, the degree of competition in British retailing may be reexamined, and the regulatory environment in which the major grocery retailers operate may well be tightened to counteract the suspicion that British consumers may be paying dearly for the success story and growing dominance of the major grocery corporations.

175

At the watershed?

... The spectacular performances of the British grocery retailing corporations in the 1980s had given rise to a certain fragility of position. In particular, the capital markets had come to expect and discount continued substantial annual increases in turnover and profit; annual increases which were intrinsically dependent upon the new-store expansion programmes. I warned in 1987 that 'failure to meet these exacting annual new-store expansion targets can have a marked effect on year-end results, and any slip will be compounded by failure to meet unrealistically demanding stock market expectations'. Four years later, Asda is clearly paying the price of such failure. Moreover, with the top three corporations now spending £2 billion per annum on their store expansion programmes and planning to increase their floorspace by 8 or 9% per year for the foreseeable future, that is, a planned doubling of the physical capacity of the market leaders by the year 2000, this potential fragility of position looks no less stark. And it is interesting to note that this view is increasingly being echoed elsewhere. For example, reviewing the June 1991 Sainsbury rights issue, the *Financial Times* (19 June 1991, page 22) raised the question of 'how long the likes of Sainsbury, Tesco and Argyll can continue to reap attractive returns from their enormous investment programmes' and ventured the opinion that:

> The trouble about food retailers is not the immediate outlook; it is rather the way all possible good news is taken for granted. One of these days, it must surely all go wrong.

So, the question must be raised – do the 1991 rights issues mark a watershed in the 'golden age' of British grocery retailing, and will that 'golden age' continue during the 1990s? The answer I believe is inevitably both yes and no.

In the simple sense of the impact of the major grocery corporations on the built environment of Britain, then clearly the 'golden age' will very visibly be seen to be continuing. Sainsbury claims to have identified an additional 160 sites for development in the medium term, Tesco's development programme is somewhat larger, and Argyll's future depends critically upon its ability to roll out its Safeway superstore format as rapidly as possible. This implies the construction of a further 350 to 400 new grocery superstores, each with an average floorspace of 35 000 ft^2, on the edges of Britain's towns and cities over the next five years. Hardly something which will go unnoticed in a built environment already transformed by the superstore revolution of the past decade, and enough to keep urban planners in Britain fully occupied. Market saturation, that hoary chestnut of British retail analysis, appears as far away as ever. Sainsbury, for example, is scarcely represented in Scotland and thinly represented in the north of England. Moreover, the quality of offering provided by the new stores is continuously being upgraded, and established locations can be revisited with a new and more profitable configuration.

But clearly, as we have seen, it will be a differently structured 'golden age'. The rights issues of 1991 suggest that the era of the 'big five' may have passed and the early 1990s will be that of the 'big three'. Moreover, it will be an age in which the locus of competition between the major corporations has once again been subtly shifted. And being located ever more firmly in the capital markets it brings enhanced risks. Net profit increases and shareholder dividend increases of more than 20% per annum may be much harder to sustain. For example, the impressive productivity gains achieved by Sainsbury and Tesco in the 1980s (a 12% real-terms increase in labour productivity by Sainsbury in 1983–87, following a 14% real increase in the previous five years) were achieved on the back of heavy investment in point-of-sale scanning equipment, computerised inventory control, distribution, accounting and management information systems, and increasingly skilled utilisation of flexible and contingent labour. These gains will be much harder to sustain in the 1990s, and it is interesting to note that Sainsbury increased net pretax profits by only 14.8% in the year ending March 1991; the first time for more than a decade that the annual rise in such profits has fallen below 20%. Moreover, there is a growing feeling that a different political climate in the 1990s might easily witness a tightened regulatory environment, and that the question of who has paid for the 'golden age' of British grocery retailing might finally have to be answered. But even if that question does not have to be faced for some time yet, net profit margins are surely at rather exposed levels.

Set against these factors is the capacity of the top grocery corporations to diversify. This will involve adding or further developing product areas such as toiletries, stationery, newspapers, flowers, and petrol in Britain (where Sainsbury already has 2% of the UK petrol market) and/or possible enhanced internationalisation of operations. In this context, Sainsbury plans to add eight new Shaws supermarkets to its US chain. The growing success of its operations in the USA reinforces the statement which I made in my 1989 commentary that the aim of such internationalisation is 'to establish sound bases with relatively straightforward growth prospects both for sales and for profits; bases which can be used *defensively* to provide opportunities for capital investment in an era in the mid- 1990s . . . when UK-based profits growth at the levels which the stock market has come to expect and discount may become more difficult to sustain' (page 288).

In the final analysis, however, it is the hugely increased funding requirements of the major grocery corporations to finance their expansionary ambitions – so vividly encapsulated in the 1991 rights issues – which prompts the question about the watershed in the 'golden age' of British grocery retailing. There is already evidence that the return on capital in the grocery retailing sector as a whole is falling in the United Kingdom. So far that fall does not appear to have affected the likes of Tesco and Sainsbury to any marked extent. But the suspicion remains that it may soon begin to do so. And certainly, it would be a brave forecaster who extrapolated the 1980s rate of real increase in capital investment by the major grocery corporations into the second half of the 1990s. The 1991 rights issues will in retrospect, I believe, be seen as an appropriate moment to have taken stock of the 'golden age' of British grocery retailing and to have begun to anticipate some of the new themes which will dominate the trajectory of this key sector of the British service economy as we move into the 21st Century.

Modernity and post-modernity in the retail landscape

J Goss, 1992 (In Anderson K, Gale F (eds) *Inventing places: studies in cultural geography*, pp. 158–77)

Introduction

There is widespread recognition of a profound shift of cultural sensibility in western societies over the last two decades or so. The various terms used to describe the 'New Times' – 'post-industrial society', 'information society', 'post-Fordism', and 'post-modernism' – suggest that it is complex and encompasses various dimensions of everyday life. This chapter will briefly examine the relationship between the material and symbolic – or more loosely, the economic and cultural – components of the ongoing transformation and its manifestation in the built environment, particularly in purpose-built places of consumption, or shopping centres or malls. This is appropriate given the importance of consumption to our contemporary lives, and given the fact that it is in architecture that the new sensibility attains its most visible expression (Jameson 1984, 54; Sharrett 1989, 162).

First, without mass-produced consumer goods everyday life would be inconceivable for most of us, not only because they sustain our material living standards, but also because they help define individual and collective identities. In the consumer society you are 'what you buy' as much as 'what you do' and, as media constantly inform us, self-actualisation is only the next purchase away. Shopping now may be the second most important cultural activity in North America, and although watching television is the first, much television programming promotes shopping directly (through advertising) and indirectly (through depiction of consumer lifestyles). The existential significance of shopping is clearly recognised in popular culture by bumper stickers shouting slogans such as: 'Born to Shop', 'Shop 'Til You Drop', and 'I Shop Therefore I Am'.

Second, the built environment reflects material and symbolic changes in society: 'Architecture is the will of the epoch translated into space' (Mies van der Rohe 1926, cited in Frampton 1983, 40). For example, the development of the skyscraper in the nineteenth century was based on advances in technology (the elevator and structural steel), the organisation of production (mergers and the rapid growth of the corporation), and prevalent ideology (verticality symbolises corporate power and classical styles symbolise the civic function of business). The built environment, however, does not merely mirror historical change, for social relations and ideologies are partly reproduced through it. The argument of this chapter, then, is that material and symbolic transformations characterising the 'New Times' are manifest and at least partially realised in the retail built environment.

The history of the shopping mall

The planned shopping centre had humble beginnings before the Second World War: the first in the United States was built in Lake Forest, Chicago in 1916: in 1922 Country Club Plaza, a prototype shopping district with stylised architecture, landscaping, unified management and a sign control was opened in Kansas City; and in 1931 Highland Park Shopping Village, the first centre based on a pedestrian mall, was built near Dallas. Other small centres were built at busy intersections, but the department stores generally remained downtown until the massive highway construction and residential suburbanisation of the 1950s. In 1950 there were less than 100 shopping centres in the United States (Urban Land Institute 1985, 16): today there are more shopping centres than post offices or secondary schools (Stoffel 1988). Nearly 35 000 shopping centres offer a total of almost 235 km² (2.5 billion ft²) of gross leasable retail space (National Research Bureau 1990), and huge super-regional malls sprawl around suburban highway interchanges and squeeze into the decaying fabric of downtown. The largest shopping centre in the world is the massive West Edmonton Mall in Canada which is 1.5 km (1 mile) long, covers about 145 ha (110 acres), has a total floorspace of 483 000 m² (5.2 million ft²) and parking for 14 000 cars.

The shopping centre is a place in which retailers sell and consumers shop, but it is more than that – it also provides for entertainment, edification, education and sustenance. It typically houses funfairs and fashion shows; hosts community dances and concerts; conducts fitness classes and courses in adult literacy: and provides food and drink. It is also a predictable, safe and sanitised alternative to the old city street, a place where families go on outings, old people idle and exercise, and teenagers hang out and grow up 'mall-wise'. The geographical spread of this cultural institution and its way of life has truly resulted in 'The Malling of America' (Kowinski 1985) and with the global export of the model we are perhaps witnessing 'The malling of the world'! . . .

The culture of consumption

Consumer goods serve the double purpose of satisfying socially defined needs and 'materialising' cultural distinctions, providing a code that symbolically expresses personal and social difference (Sahlins 1976; McCracken 1988). This is not entirely new, as 'consumer culture' has been in the making for several hundreds of years (Braudel 1967), and the first consumer revolution, which took place in the nineteenth century, had already established consumer goods as repositories of social meaning (Williams 1982; Miller 1981). It is only in this century, however, and particularly since the Second World War, that everyday life has been so thoroughly commodified that we can be persuaded to buy sexual attraction, happiness and personality, as well as status, in the form of consumer goods. Persuasion is the responsibility of specialists in the cultural institutions of advertising, marketing and the media. Identified as the 'captains of consciousness' (Ewen 1976, 19) these agents might better be called the 'high priests of capitalism' since their means of persuasion is summed up by the notion of 'fetishism of the commodity'.

A fetish is an object of religion in 'primitive' cultures that is invested with spiritual powers and regarded with dread or reverence. Commodities work similar magic in contemporary western cultures such that advertising is 'a highly organised and professional system of magical inducements and satisfactions, functionally very similar to magical systems in simpler societies, but rather strangely coexistent with a highly developed scientific technology' (Williams 1980, 185). Highly sophisticated advertising techniques maintain the superstition that possession of the physical object confers power over nature and others even if the 'real' power lies in the economic or political capacity of the owner.

A second sense in which the commodity is fetishised is in the 'masking' of the social relations necessary to produce it and the human labour it embodies. Although some commodities are marketed as the products of 'craft' labour (signifying quality), sensitive contemporary consumers would generally rather not be reminded of (or haunted by) the third world sweatshop labour that makes their designer clothes or the assembly lines that produce their household goods. Commodities appear in advertisements and the marketplace with the ghost of human labour thoroughly exorcised, so that very few consumers know, or can give thought to, what they are composed of, where they were made and who made them (Jhally 1987, 49).

However, one must not accord too much power to either the magician or to the advertiser, for their operations only work for an audience predisposed to believe in the illusion, wherein lies the real source of the 'magic' (Bourdieu 1986, 137). Designers, advertisers and retailers do not have to conspire consciously to deceive their audiences (although they often do), but may merely highlight 'latent correspondences' between the commodity and cultural symbols (Sahlins 1976, 217). It is not necessary to tell us that fast cars confer extra libido upon drivers, or that a particular cigarette brand will add 'cool' to the smoker. We take the attractive (female) passenger, or the elegant decor seen with the commodity and 'independently' make the connection. Consumers are asked to employ their accumulated cultural knowledge to actively weave together the natural, symbolic and social elements provided by the image-maker and so create the commodity's context – that is, the mode and manner of its consumption (Sack 1988). Moreover, consumers are never only the dupes that the 'captains of consciousness' might wish them to be. They may actively and imaginatively subvert the images presented and latent associations highlighted, perhaps by 'unmasking' the social relations embodied in the commodity, by consciously consuming out of context, or campaigning in general against the manipulative content of imagery.

Finally, a critical component of the commodity's context is the real or imagined landscape in which it is advertised or marketed. The advertiser employs the power of place to suggest the appropriate mode of consumption. The consumer then imaginatively employs the commodity to locate him/herself in this place and to weave the appropriate context. The tropical beach, for example, symbolises relaxed, sensual luxury, and the consumer is able to experience this through the consumption of particular clothes, cigarettes or alcohol, while lounging on the back porch in a suburban subdivision. . . .

Post-modernity and individualised consumption

Under the 'post-Fordist' regime of accumulation, mass production and consumption have given way to flexible production and personalised (as opposed to mass) consumption. Sophisticated production technologies and computerised distribution systems allow rapid turnover of product styles designed for the specific market segments identified and exhaustingly researched by the new 'disciplines' of geo-demographics and psychographics. The demand for high-quality information and sophisticated co-ordination has increased the number of specialists in the culture, knowledge and communication industries who provide services essential to commodity production and circulation. These specialists form the core of what has been called a 'new middle class' and their lifestyles and cultural orientations are critical to the consumption practices of post-modernism.

The new class is defined as a waged class because it does not own the means of production nor the product of its labour, yet has a degree of control over the production process and may claim intellectual ownership of the product. It is thus in a somewhat ambiguous class position, but because it is relatively well-educated, it is able to employ cultivated distinctions in taste, lifestyle, personal expression, sexuality and quality of living environment to define its cultural territory. These distinctions are expressed in consumption and are most readily employed by the subgroup of this class popularly known as 'yuppies'. They are stereotypically associated with commodities that, for example, exhibit cosmopolitanism (from Japanese paper lanterns to espresso coffee machines), eschew ostentation (minimalist furniture and natural finishes), boast quality (brand names and designer labels), and display privileged knowledge (gourmet coffees and the 'right' wines).

Class distinction is thus no longer quantitative (based on the value of commodities consumed) but qualitative (based on style of consumption). In fact, one might say that it is not primarily the material object that is consumed, but the image of ourselves consuming the object (Baudrillard 1981). One no longer buys merely to 'keep up with the Joneses', but to appropriate a style for one's persona. Thus even with a practical household gizmo one gets literally 'The Sharper Image'; and with fashionable clothes one gets 'The Look', becomes a member of 'The Limited', or is seen to be on the right side of 'The Gap'. Ironically, however, as entrepreneurs expand the market behind the cultural avant garde and the 'masses' emulate the middle class by consuming its commodities, so the new class must develop new tastes to mark its distinction. The last fad is rapidly replaced with the latest, making for an extraordinarily rapid turnover of symbolic content in commodified experiences of tourism, leisure, sport, entertainment or body maintenance. Identity is dynamic and emphasis is on self-discovery and personal growth through self-improvement literature, personal improvement seminars and image consultants.

Disillusionment with the political failures of the 1960s and the rise of this 'culture of narcissism' (Lasch 1979) have translated societal problems into personal inadequacies, social concern into self-help, and public life (of the festival, voting and community) into appearance. Immediate personal gratification is pursued instead of life-long co-operative projects such as marriage and child-rearing, and pursuit of quick financial success replaces the career goal. Ironically, however, having been liberated from the constraints of these institutional projects post-modern individuals yearn for a sense of their history and place. Nostalgia manifests the post-modern desire for authenticity, for the continuity of tradition and for community lost. In an existential search for roots (Jager 1986) the post-modern voraciously consumes styles of past times manufactured by the 'heritage industry' (Hewison 1987) and of distant places produced by the tourist industry.

The past is commodified in 'pop images and stereotypes' (Jameson 1983, 118), such as old 'B' movies, retro clothing, and restored pinball and soft-drink machines. Historical artifacts are 'museumised', while the contemporary other is ransacked for signs of tradition and community. Consequently, the post-modern consumer has accumulated a 'well-stocked musée imaginaire' (Jencks 1987, 95), or fragments of experience from other times and places: (s)he has typically learned to eat muesli for breakfast and Ethiopian for lunch, drink Mexican beer and Chilean wine, wear Red Army surplus and sarongs, watch subtitled movies, listen to reggae, rai, opera and gamelan, and dance the hula or lambada. These souvenirs and disembodied gestures mark the cosmopolitan lifestyle and sophisticated taste of an individual, and show an obsessive desire for authenticity in an increasingly rootless culture. . . .

The post-modern retail environment

The essential forms of the post-modern retail environment – the specialty centre and the downtown 'megastructure' – reflect the vernacular and high forms of post-modernism respectively, while a hybrid form – the festival marketplace – combines elements of both. The specialty centre is an 'anchorless' collection of upmarket shops and restaurants pursuing a specific retail and architectural theme. It is prone to quaintification. Typical designs in North America include New England villages (Pickering Wharf, Salem, Massachusetts): French provincial towns (The Continent, Columbus, Ohio): Spanish-American haciendas (The Pruneyard, San Jose): Mediterranean villages (Atrium Court, Newport Beach, California): and timber mining camps (Jack London Village, Oakland, California). Pride of place must, however, go to The Borgota in Scottsdale, Arizona, a mock thirteenth-century walled Italian village, with bricks imported from Rome and shop signs in Italian (Kowinski 1985, 233), and to The Mercado in nearby Phoenix, Arizona, modelled on traditional hillside villages of Mexico, with original components imported from Guadalajara, and buildings given Hispanic names.

The details may be so accurate that authenticity is displaced and the stylised copy appears more real than the original. The restaurants with their architectural elements, textual fragments, objects of material culture attached to walls and ceilings, and of course the modified cuisine may be more convincingly Italian or Mexican than those in Italy or Mexico. The consumer ignores or forgets that the tortillas are factory-made, the entrée warmed in a microwave oven, and the decorative handicrafts probably produced for export under elaborate subcontracting systems. While the consumer is presented an opportunity to display acquired exotic tastes, the commodity on sale is perfectly fetishised.

The downtown megastructure, on the other hand, is a self-contained complex including retail functions, hotels, offices, restaurants, entertainment, health centres and luxury apartments. Typical examples include Water Tower Place in Chicago, the Tower City Centre in Cleveland and Town Square in St Paul. These structures are part of the challenge to the separations (between moments of production, reproduction and consumption, and between workplace, living place and leisure space) on which modernist culture is founded. At the same time, these small worlds ensure that the needs of affluent residents, office workers, conference attenders and tourists can be met entirely within a single hermetically sealed space.

Several features distinguish the downtown megastructure from the suburban shopping mall, although by now many of these have been extensively 'retrofitted' in the post-modern style. After studies found that 70 per cent of all energy consumed in malls was spent on lighting, the calculus of economics and fashion have combined to return daylight in glazed malls reminiscent of nineteenth-century European arcades. These afford a sense of grand public space, and natural light allows the planting of ficus, bamboos and 'interiorised' palms to simulate the tropical environments of tourism, to indicate a respectable age for the establishment and to suggest care for the environment. Water has always been an important element in the mall as a means of soothing tensions and refreshing shoppers, but now elaborate watercourses and waterfalls simulate nature, rather than urban fountains. These three effects combine to turn things inside-out, so that pure and perfected nature, the ideal place of leisure, is ironically found indoors within the city, and no longer in the deteriorating environments of the suburbs beyond.

The downtown malls are also no longer primarily 'machines for shopping', although the aesthetics of movement are retained in the sweep of huge escalators and the trajectory of 'bubble' elevators. Now passage through the mall is an interactive experience, an adventure in winding alleys resembling the Arabian souk or medieval town, with the unpredictability of 'pop-out' shop fronts – glass display cases which jut out into the mall – and mobile vendors. Shopping at the downtown mall is not merely the necessary purchase of goods, but is a form of retail tourism where the individual makes her/his own itinerary. . . .

No mall experience is complete without food, which has become a critical marker of social taste. Food courts and full-service restaurants now offer not merely sustenance to the hungry shopper, but a full range of culinary experiences, from fast food to five star, and from international to local ethnic cuisine. The food court is also a place to rest and a vantage point from which to view the spectacle of the new middle class at play.

Entertainment is more than ever the key to success, and attractions such as ice rinks, carousels, roller-coasters, local and historical exhibits, and staged events are an essential part of the show. The contemporary retail environment is an exercise in Disney's 'imagineering', the employment of fantasy and engineering technology (Relph 1987, 129) which effectively enlivens, or conceals, the practical activity of shopping. Also significant in this regard is the fact that shopping centres are increasingly graced by 'high' cultural activities. Developers have commissioned artists to create special works integrated into the design of the centre, established valuable collections of artists' works in permanent displays, sponsored temporary exhibits and hosted shows of classical music. For example, the Bel Canto opera competition is held in shopping centres across the country; a Shakespearean Festival is held in Lakeforest Mall, Gaithersburg, Maryland; sculptures by Henry Moore and Jonathan Borofsky are exhibited in 'sculpture courts' at North Park Center, Dallas; South Coast Plaza in Costa Mesa. California boasts 'one of the most important outdoor sculpture environments in the world' (*Shopping Center Age* 1989, 108); and South Coast Plaza, Faneuil Hall Marketplace in Boston, The Mercado in Phoenix and Horton Plaza in San Diego all have art centres or museums on the premises. This aestheticisation of the shopping experience enables the new middle class to develop and display its cultural capital while obviating charges of vulgarity associated with conspicuous consumption.

The festival market combines these elements with an idealised version of historical urban community and the street market, typically in a restored waterfront district after the model of Faneuil Hall Marketplace in Boston. These environments reflect a nostalgia for manual labour, public gatherings and the age of commerce. Buildings and vessels are restored, and there is usually a historic museum on site. The marketplace is typically decorated with antique signage and props which casually suggest an authentic stage upon which the modern consumer can act out a little bit of history. The street entertainers, barrow vendors and costumed staff often support this stylised image.

The aestheticisation and historicisation of shopping is appropriate because an increasing amount of cultural education is required to appreciate commodities. As both the audiences and the techniques of the 'captains of consciousness' (Ewen 1988) have become more sophisticated, the cultural symbols employed in advertising are more complex. The young professional selectively shopping for quality goods in specialist boutiques is employing perhaps as much accumulated cultural knowledge in creating the context of the commodity as (s)he would in interpreting an oil painting, theatre performance or historical novel. Through design strategies employed in the post-modern shopping centre the act of consumption itself is fetishised: the material activity of shopping resonates with the symbolic activity of leisure, entertainment, education and artistic appreciation.

Conclusion

The history of the planned built environment of retailing is more complex than this brief sketch can shown, but in general the unitary 'shopping machine' of the suburbs is being replaced by the mixed-use 'consumer spectacle' originating in the city. This shift in form and function illustrates and reproduces some of the economic and cultural determinations of the 'New Times', a more or less fundamental transformation in the nature of western capitalism. Consumption has replaced production as the driving force of social life. The dynamic connection between the cultural and economic dimensions of existence, however, ensures that the increased significance of consumption and the symbolic ordering of social life is linked to profound changes in the nature of production and the material basis of existence. The argument that has been made here is that the built environment is a particularly useful social object through which to explore the linkages between them.

CHAPTER 8

Public sector restructuring

The significance of public services

> Service economics are dominantly only considered from the angle of the
> profit-making sector ... [yet] ... government activities gather the majority of
> service employment. (*Lettre de Liason des Service*, 1990: 3)

The growth of employment in private business and financial services
has received most attention in studies of service location, while in
contrast public services have been relatively neglected. Yet public
services, delivered by various combinations of local and national
organizations, cover a wide range of activities, including education, tax
collection, defence, law and order, health care, sanitation, and
emergency and social services. Some operate alongside an often much
smaller private sector, which offers complementary or competing
provision, for example in education, health and even in security. The
principle that all citizens should have access to such services cannot,
however, usually be satisfied through market forms of operation. Some
groups and places are always likely to be unprofitable to serve. Public
service effectiveness tends therefore to be judged by social or political
criteria, including equity of access and quality of delivery, rather than
by purely economic criteria of cost or profit.

There are other reasons why services are provided by state
organizations. Some are a 'pure public good' which cannot generally be
provided by individuals or the private sector. These include law and
order, defence, and the administrative and regulatory functions of the
state, such as tax collection, local planning and international relations.
In many countries also 'natural' monopolies are publicly owned,
especially those based on the control of infrastructure networks such as
telecommunications or rail transport. Benefits are often claimed from
the coordination and planning of such basic investment, and public
control is also sometimes justified by national security considerations.
These claims have been challenged in recent years, however, resulting in

184

a widespread trend of 'deregulation' or 'privatization' of such activities to attract private investment and ownership, for example in tele-communications, energy and water supply, rail and air transport.

Most resources for public services are levied nationally, through various forms of direct and indirect taxation, but they are typically delivered to meet diverse local needs. Their organization is thus commonly hierarchical, or classically bureaucratic. The balance between central and local influence or control varies between services; some are provided entirely by central government agencies, while others are delivered by various levels of regional or local government, although often still under central direction or regulation.

The public sector (which as defined by OECD is largely public services) makes a significant contribution to the economy; accounting for between 6 and 33 per cent of total employment in OECD countries (Table 8.1). Japan and to a lesser extent the US are exceptional; they have a small public sector, and increases in public sector employment have been less pronounced. The expansion of the public sector has also been matched by employment growth in the private sector. European countries, in contrast, have experienced a substantial growth in the share of public sector employment in the economy; for example in Germany it rose from 8 per cent in 1960 to 16 per cent in 1988, and it increased from 15 to 21 per cent in the UK over the same period. The majority of governments regard this rapid growth of the public sector as harmful to their economic prospects. During the 1980s, most set about restraining growth. Even so, in Europe, though public sector employment growth virtually stopped in the 1980s, there was little change in its relative contribution to the economy, because of slow employment growth in the private sector (Table 8.1).

The reasons for the growth in public service employment in the advanced economies include the following:

Table 8.1. Public sector employment relative to employed labour force in OECD countries

Country	1960	1972	1981	1988	Peak % share Value	Year
US	14.7	16.5	15.3	14.4	16.8	1975
Japan	na	6.0	6.7	6.3	6.7	1981
Belgium	12.2	14.4	19.3	20.4	20.8	1986
Denmark	na	20.2	29.8	29.8	31.0	1983
Finland	7.7	12.6	18.4	21.7	21.7	1988
France	16.8	17.7	20.5	23.0	23.1	1987
Germany	8.0	12.2	15.2	16.0	16.1	1987
Italy	8.7	13.1	14.8	15.5	15.5	1988
Norway	na	20.8	23.9	25.8	25.8	1988
Portugal	3.9	8.6	10.3	12.8	12.8	1986
Sweden	12.8	23.1	31.5	31.7	33.1	1985
UK	14.8	19.4	21.8	21.2	22.4	1983

Source: Jefferson and Trainor (1993).

1. Demands for improved health care and education services.
2. The expansion of public transport and communications infra-structure.
3. The growth of defence spending, until recently, reflecting the 'Cold War' between East and West.
4. Demographic changes, including growth in the numbers of elderly in the population, with greater needs for health and welfare services.
5. Changes in family composition, including an increase in the numbers of women working, and growth in the numbers of single parents, which add to the demand for child care and welfare services.
6. Interventionist policies during the 1960s and 1970s, to guide economic growth and restructuring, requiring more government workers.
7. Greater physical planning controls to restrain or guide urban and industrial growth, thus minimizing its environmental impacts.
8. Improvement in all parts of the public sector was seen to be an integral part of the long post-1945 economic boom up to the mid-1970s.
9. More recently, growing unemployment.

Such were the needs, especially of the expanding welfare state, that the efficiency of public services in spending the resources raised from taxation and other levies was generally not subject to systematic scrutiny. However, the economic slow-down of the 1970s and 1980s, and concern over the inflationary impacts of rising public spending led to a drive for cuts in some public services, and greater efficiency in providing others (Ladd, 1992). Such pressures have been compounded by the end of the Cold War which had justified high levels of defence spending (Lovering, 1985; Breheny, 1988; Lovering and Boddy, 1988) One consequence has been the progressive reorganization of much public sector provision (Pinch, 1989; Mohan, 1992). Local government in the UK, for example, came under stronger spending control from the centre, while nationally supervised management agencies assumed responsibility for delivering many services (Box 8.1). The balance of provision also shifted in some sectors from public agencies to private or voluntary sector organizations (Box 8.2). Such changes have reduced the numbers of public service workers in many departments, especially in the major urban centres where they are concentrated, but at the same time, demand-led growth has continued to produce pressures for expansion.

The geography of public services

The location of public services can be quite different from that of the private sector. Welfare goals suggest that, ideally, people should have equal access to social and health care services wherever they live (Smith,

Box 8.1:   Reorganization of the UK central government civil service

The reading by Winckler covers policy and locational changes in the UK central government civil service between 1962 and 1985. Since then there has been a further significant reorganization. The UK government's White Paper 'Competing for Quality' (1991; CMND 1730) summarizes recent policy towards the public sector:

> [Introducing] greater competition . . . has gone hand in hand with fundamental management reform. . . . This means moving away from the traditional pyramid structure of public sector management. The defects of the old approach have been widely recognised: excessively long lines of management with blurred responsibility and accountability: lack of incentives to initiative and innovation: a culture that was more often concerned with procedures than performance. . . . [P]ublic services will increasingly move to a culture where relationships are contractual rather than bureaucratic.

Reflecting these changes, by 1993 civil service employment had declined to 565 000 jobs from 732 000 in 1979; the bulk of these losses being experienced in large administrative centres: London, Newcastle, Edinburgh and Manchester. Tight financial and manpower control helped reduce civil service numbers. Where possible, functions were sold to the private sector (Box 8.2). This markedly reduced the scale of the Department of Defence, where support functions such as the Royal Ordnance factories and Royal Navy dockyards were privatized. But until recently the bulk of the civil service, which supplied the core of the welfare state, remained relatively untouched.

However, in 1988 the civil service was split into two (Cabinet Office, 1988). A core section continued to advise ministers and was responsible for managing contracts, but large parts of the service were divided up into quasi-independent agencies responsible for delivering particular services. By February 1994, 94 agencies employing approximately 60 per cent of the civil service had been established, and another 21 per cent of civil service employees were under consideration for agency status.

The division of the service into separate agencies has encouraged the view that a further round of privatization may be possible (Mather, 1992). Already drawing on previous policies towards local authorities, agencies have been forced to contract out services to the private sector (Department of the Environment, 1985; Moore and Parnell, 1986). If this policy direction is maintained, it will be possible to talk about a 'minimalist' state, in which the number of state employees is reduced dramatically, though government budgets remain substantial. The delivery of services is largely left to the private or quasi-private sector, with the public sector retaining responsibility for setting the terms of contracts and monitoring compliance (Ridley, 1988).

Observers believe reforms are fragmenting government. Corby identifies 'a clear trend towards breaking up the unified civil service' (Corby, 1991: 39). More critical commentators speak of the 'dismemberment of government' (Phillips, 1992: 21). Greater flexibility is being introduced so that agencies can attract and financially reward staff. Effectively this means that agencies have the freedom to pay staff more in the South East and London, where they have the greatest difficulties in attracting staff (Starks, 1991). More speculatively, as the services provided by the civil service become more diverse and fragmented, involving a mix of private and public provision, and varying employment and conditions of service, the civil service may cease to operate so effectively as a counterweight to private sector centralization. Government functions, whether ultimately performed by the public or private sector, may provide employment similar to the local labour market in which they are based.

Box 8.2: The privatization of telecommunications

During the 1980s, in response to similar trends in the US, the UK government 'privatized' and 'liberalized' the telecommunications industry, hitherto run as a public sector monopoly (Noam, 1992). BT (British Telecommunications) was established as a private company, and Mercury was allowed to construct an alternative telecommunications network. BT's monopoly on telecommunications equipment was broken, and private operators were able to offer new advanced telecommunications services over BT's network (Marshall, forthcoming). Subsequently, cable television companies have been given permission to provide telecommunications services. This transformation and its geographical impacts raises important general questions about the spatial consequences of introducing competition into public services.

As a state monopoly BT provided a national telecommunications service. Telecommunications investment in rural and peripheral areas, which have little traffic, was cross-subsidized by profits earned on long-distance trunk and international routes. This did not mean that new investment in networks and services was introduced everywhere at the same time, but there was a commitment to 'universality' of service provision.

In a more competitive environment, BT has been forced to take a more commercial attitude to the operation of its telecommunications network, with prices for telecommunications services moving closer to cost. Its network competitors, Mercury and more latterly cable TV companies (which can also provide telecoms services), have naturally tried to serve what they believe to be the most profitable markets, which tend to lie in larger metropolitan areas with high demand for telecommunications. BT has responded to this by offering its new advanced services in these areas first.

So far the effects of privatization and liberalization seem to be following the US pattern, which has a longer history of liberalized telecommunications. As Muligan argues:

> In the UK and US . . . [telecommunications] are evolving into a complex patchwork of private networks, LANS (local area networks), overlays, microwave towers and satellite transponders and cable television nets. . . . Each special purpose network . . . built in response to demonstrable demand. (Muligan, 1991: 226)

This means that the government's competitive approach to telecommunications has an uneven spatial impact. In areas of lower demand for telecommunications, users remain reliant on BT, while in areas of higher demand users have much more choice of networks and services. The 'universality' of telecommunications has been undermined (Gillespie and Robins, 1991).

1976). In some cases, it has been legally stipulated that at least minimum standards are widely available. Thus many public services must be provided relatively locally in some form, unlike many private services aimed at profiting from the better off and more mobile social groups. Government may also choose to bear the costs of operating services in marginal locations to encourage their economic development. The widespread distribution of public services can thus act as a counterweight to the growing concentration of private services in and around the larger urban areas.

Though traditional policies of geographical dispersion of provision and subsidized access are designed to overcome the worst inequities associated with location, in practice inequities in accessibility to public services are inherent in geography, arising at the local, urban and interregional scales for different services (Pinch, 1985; Steaheli, 1989). At a local level, small communities may support a doctor's surgery, a primary school and a post office, with clients having to travel only short, although still unequal, distances for their basic needs. More capital-intensive and specialized services, such as hospitals, secondary and tertiary education, and local government agencies, have always been located in towns and cities. It is assumed that some people will travel farther than others for such services, which need a large population base to support their high costs. In some cases, public transport or ambulance provision enables those living in remote settlements, or needy groups such as the aged, to travel free or cheaply to such centres. Some mobile services, such as libraries or health clinics, may even travel out to them.

Such disparities are compounded by government policies towards the public sector which take little account of geography. The uneven spatial impact of national defence expenditure is well documented (Lovering, 1985), but a range of other national policies have unforeseen geographical outcomes, through their impact on the location of government employees who must carry them out (Marshall, 1990). Even the geography of health care, not normally regarded as having an uneven spatial distribution, is a product of a plethora of national and local administrative decisions, including privatization and resource redistribution (Mohan, 1988a). Many services also combine public and private forms of provision, or have been initially provided by private operators, and subsequently brought into the public sector. Their geography will reflect this.

Several trends over the past two decades have particularly affected patterns of public service provision (see also Table 4.1). As a result, such provision has become more concentrated in and around larger centres, cutting out small-scale local facilities:

1. All public sector costs have come under close scrutiny, with stricter judgements being made about how many and which people benefit from particular spending programmes.
2. The growing automobile-based mobility, especially of the more prosperous sections of the population, has enabled them to travel more easily to major centres. This has reduced demands for special and public transport provision, making them vulnerable to cost-saving cuts.
3. The capital and organizational costs of providing specialist services such as hospital treatment, education or local government have risen, favouring their concentration into fewer centres, in progressively larger urban areas. This trend increases the need for people to travel farther, even while public transport provision declines.

Where the income gap between better-off and poorer groups has grown since the 1970s, it has been further widened by the geographical reorganization of many public services. While improving the range, and even reducing the relative cost of specialist public services for many, reorganization has also made them less accessible for vulnerable groups, including those without cars, such as older people, poorer families, especially mothers, and those living in rural areas (Pinch, 1985). Village and small town schools have been closed, group practices have replaced the individual doctor, and small post offices are threatened by closure. Public services thus follow trends in many commercial services. The degree to which these can be balanced by continuing to subsidize local provision or selective mobility remains a matter of local policy, but this is increasingly implemented against a background of political pressures to reduce tax-supported costs.

Urban inequities

Another arena of public service change has also become the focus of debate in many countries; their withdrawal from inner city areas. Public services are generally concentrated in inner cities, the legacy of past investment in the most populous and accessible centres for health, education and public administration. They were also often surrounded by the most needy areas of inner urban deprivation. This inheritance has been challenged by the following:

1. Counter-urbanization trends, spreading the demand for services more widely, and reducing the accessibility of more of the population to inner city areas.
2. The reduced local tax base for supporting urban-based public services, increasing their dependence on national or state subsidies, especially as the better-off move out. This affected US cities first, where counterurbanization began earliest and is most dispersed, and where suburban communities are administratively and financially independent. The same tendency is evident in Europe, however, with dispersal following US patterns.
3. The high cost of public service provision in inner city areas. Wage, maintenance and transport costs are high, while many old buildings are expensive and difficult to adapt to modern needs. Thus, in line with what has happened to private services such as retailing and some office functions, the pressure to reduce costs in public provision favours relocation away from inner cities.
4. The reduced accessibility of inner cities even to many needy areas, including increasingly some of their own poorer suburban and outer housing areas. Problems of traffic congestion create chronic problems especially for operating emergency services.

Restructuring, including a reduction of inner city public sector provision is therefore to some degree a rational response to changing needs. It carries significant social implications, however, which are matters of wide political debate, especially over health care and social service provision (Mohan, 1988a). Inner cities remain centres of high demand for public services, whether from surrounding poor neighbourhoods, or from those who may continually move in from elsewhere. High levels of urban provision are still needed, therefore, as part of general social support to the poor. Further, employment in such areas is often heavily dependent on the public sector. Large-scale reductions in public employment thus particularly affect inner city economies already undermined by private sector disinvestment. Any public service change requires political sensitivity to the competing priorities of efficiency of resource use and maintaining equity of access. The complexity of these issues, however, is most intense in the inner cities. Here, the general dilemma of balancing the interests of (a) the better-off and the poor and, (b) the taxpayer and the needy, must also engage the interests of (c) old parts of cities and of the suburbs and beyond, and (d) public service workers as well as public sector consumers.

Regional inequities

There is an even greater scale of interregional inequity in public service provision, for functions serving nationwide needs. Winckler (see reading) shows that many central government departments need to be in the capital, close to the politicians responsible for them, where they may in turn attract the head offices of large firms, financial services and higher-order business services. Means have nevertheless been sought in many countries to decentralize some government functions, to reduce the costs of service provision, to provide employment in areas with economic problems and to improve the quality of life for employees living in congested metropolitan centres (Jefferson and Trainor, 1993; Pacini et al., 1993).

The administrative structure of the state has an important influence on the distribution of government activities. In countries with a tradition of strong central government, for example France or the UK, the intense capital city concentration of public services has been ameliorated in recent decades by moving out more routine clerical work and establishing new departments in problem regions with high levels of unemployment. Such government services have thus become relatively more evenly distributed across the country than many high-level private services. The continuing central control of government nevertheless limits the effect of these decentralization policies. Though savings, in terms of office and labour costs, may be achieved by decentralizing clerical work, many senior civil servants who require regular access to central functions, cannot be relocated without significant communications costs (Winckler, page 208).

Other countries have a more devolved, federal structure of government, including the US, Germany and Switzerland. Administrative functions thus have considerable regional autonomy, and may act as a stimulus for other local economic developments, a magnet for private sector activities and a powerful lobby to attract further government spending. In such circumstances, the location of central government functions is less likely to be decided purely on cost grounds, as the widespread redistribution of German government functions following the unification of East and West shows (Pacini *et al.*, 1993). Such devolved systems of government, with multiple hierarchies of administration and control are said to lie behind the more spatially decentralized nature of the economic structure of such countries (Jaeger and Durrenberger, 1991; Pacini *et al.*, 1993).

A changing context for the public sector

Different views have been taken of the modern increase in public service employment. It was possible until the 1970s to shrug off the growth in the size of the public sector as simply reflecting the need for the welfare state to catch up with past deficiencies in social policy, and for government to support private investment (Blackaby, 1979; Heald, 1983). Many, however, have since lost faith in the need for big government supplying standard services to a mass population (Stoker, 1989, 1991). They argue for a more customized, selective and entrepreneurial approach to service provision (Osborne and Gaebler, 1992; Mishra, 1984). The view has also gained ground that the public sector is 'parasitic' and a drain on the resources of the private sector. Bacon and Eltis (1978), for example, argued that investment in public services crowds out private sector investment. Such ideas struck a chord with the neo-liberal political perspective which gained power in the 1980s, especially in the US and the UK. This placed reducing taxation and public sector spending at the forefront of an economic policy designed to facilitate individual enterprise.

The increasing openness of national economies has certainly exposed the limitations of public expenditure as a means of stimulating national economic development (Morris, 1987). Public services are more likely to be viewed as a cost of production in globally competitive markets, rather than a means of maintaining domestic consumption (Jessop, 1991a). Many public services have low apparent productivity, being labour-intensive, requiring limited capital investment, and involving extensive contact with customers. As early as the 1960s and later, Baumol (1967, 1985) suggested that if wages in the public sector rose at the rate of other sectors, then 'relative costs will rise without limit . . . [and] If relative outputs are maintained an ever increasing proportion of the labor force must be channelled into these activities' (Baumol, 1967: 419–20). The implication was that the long-term growth of public services is unsustainable, certainly without many public services continuing to offer very low wages.

In Chapter 2 we argued that such generalizations fail to take account of the diversity of service activities and their contribution to other sectors of the economy. Nevertheless, the economic slow-down since the 1970s, and especially increased unemployment, have created a structural gap between a growing demand for labour-intensive public services, and the willingness or ability of governments to pay for them from taxation (Jessop *et al.*, 1991). The responses to such pressures have included not only reducing or cutting out particular services, as we have seen, but also increasing private sector involvement through contracting out particular support functions, or even returning public services to private ownership (Scarpaci, 1988; Box 8.2). Financial pressures have also generally encouraged more efficient practices, and many services have been reorganized (Jessop, 1991b). Agencies have been established at 'arms length' from government bureaucracy to introduce more commercial business practices and improve service delivery (Box 8.1). More devolved forms of management have been introduced, but with more critical supervision of their cost effectiveness (Marshall, 1990). Experiments have been undertaken with novel forms of service delivery, involving voluntary organizations or new technologies (Gershuny and Miles, 1983; Blackburn *et al.*, 1985). In general, these changes in service delivery have reduced the growth or resulted in declines in public service employment, though inevitably this has not been universal for all public services. Some have continued to grow, driven for example by increases in unemployment, demands for off-the-job training and growth in crime.

A new geography of public services? Examples from the UK

Each country has a distinctive public sector, with a different mix of public and private provision, and central or local systems of delivery. These differences reflect the political and cultural history of the country, and the administration structures which have been developed to run it. Here we examine changes in the public sector in the UK, and their geographical implications. The transformation of the public sector here has been profound, and illustrates some general lessons, especially with regard to new patterns of uneven geographical development introduced by reorganization (*cf*. Christopherson reading, Chapter 4).

Public services change in the UK has attracted active research and commentary in political science and public administration (Hennesey, 1989). These have sometimes linked the changes to the wider debates about the transition from 'Fordist' to 'post-Fordist' forms of economic organization, discussed in Chapter 4. This includes changing inter-dependence between the private sector and the state, with a reduction in classic 'welfare state' provisions, as well as general trends towards increased flexibility in public as well as private sector labour markets (Duncan and Goodwin, 1988; Jessop, 1991b; Painter, 1991; Goodwin *et al.*, 1993). Geographical interest in public services has also drawn on Doreen Massey's perspective on economic restructuring (summarized in Chapter 4, page 59).

Evidently, a restructuring approach, with its emphasis on the relationship of public policy to wider trends within capitalism, and especially on periodic reorganizations in the division of labour, is well suited to the analysis of labour-intensive public services. Both **readings,** by Pinch and Winckler (pages 198–212) highlight the significance of the public sector as an employer, especially in areas of economic disadvantage or private sector decline. They also review the geo- graphical effects of attempts to reduce costs and provide better 'value for public money' in the public services (*cf.* Table 4.1) . More still needs to be known about the impacts of restructuring on the skill, gender, age and ethnic composition of the public sector workforce, as well as on the quality and efficiency of service provision in various places. Evidence is nevertheless emerging of pronounced geographical variations in the way public services have responded to change (Boxes 8.1 and 8.2).

At the regional scale, we have seen that parts of the public sector have been used as a tool for regional development, and that areas affected by private sector decline are especially dependent on it. Such locations may therefore bear the brunt of public expenditure cuts. More generally, rationalization and reorganization may undermine the role of public services as a geographical counterweight to private sector centralization (Box 8.1). This is the conclusion of Sjoholt (1993) in relation to the reduced role of the state envisaged in Nordic countries. The range of public services described by Pinch, however, and the diverse types of restructuring being experienced, suggests that geographical impacts in the UK, at least, are likely to be more complex than this simple description suggests. They will reflect the inherited mix of departments and agencies in each area and how each is affected by different types of reorganization (Marshall and Alderman, 1991).

Restructuring has different implications for more prosperous regions. Mohan and Lee (1989) show how the growth of the private sector in London and the South East of England during the 1980s threatened the ability of the public sector to maintain services. Inflationary pressures on property and other living costs meant that teachers, nurses and other scarce staff could not be attracted by lower public sector wages. In these circumstances, new opportunities arise for private sector provision, although with a different geography from state activity (Allen, 1988b). Being dependent on consumer incomes, private services are naturally attracted to more prosperous parts of the country. Mohan (1991, 1988b) shows that private health care is focused in London, and a similar southern bias is found in private education (Bradford and Burdett, 1989). A new international division of labour has even developed in the more specialized forms of private health care, further supporting its concentration in core regions (Mohan, 1991). So the move towards private provision creates sharp geographical unevenness in the distribution of some high-level services (see also Box 8.2).

The costs of public service delivery in prosperous centres may also drive locational change in other public services. Thus, decentralization of the civil service from London to locations where office and labour costs are lower was pursued with vigour during the latter part of the 1980s. This decentralization differed from earlier relocations, from the 1960s onwards, which were planned to support economic development in problem regions as well as reduce costs. Tighter financial constraints in the 1980s, and the desire for market-based solutions, resulted in fewer short- and long-distance relocations. Long-distance decentralization to problem regions was ruled out by the increased communications costs back to London. Short-distance movement was also less likely, however, because rents and labour costs throughout the South East were not significantly below those in London. Decentralization, therefore, was predominantly to a middle band of distances, between 80 and 240 kilometres (50 and 150 miles) from London (Marshall *et al.*, 1991).

Finally, the reorganization and rationalization of public services has broken up the relatively uniform national character of many services (Boxes 8.1 and 8.2). The growth of semi-independent agencies to carry out civil service work and the use of private contractors with a different business culture have introduced greater variety into public service provision. Local pay bargaining, moving away from national agreements, adds to this diversity. Evolving differences within public services in terms and conditions of employment could mean that they will reflect local labour market conditions more, as well as become more diverse in character. In the telecommunications industry changes in pricing and patterns of investment, following privatization and liberalization, have favoured large metropolitan areas with greater demand for services (Box 8.2). How far, then, will geographical differences in public services come to reflect those in private sector services?

Conclusions

Public services, like the private sector, are subject to restructuring in response to changing economic circumstances. As the reading by Winckler points out, the government's approach to managing the economy plays an important role in shaping the character of the public service. Growing antipathy towards a large, hierarchical, public sector, providing uniform services to large numbers of the population, has led to attempts to reduce the size of and to reorganize the public sector. The case studies of the civil service and the telecommunications industry in the UK show that the introduction of competition into public services will have uneven spatial impacts (Boxes 8.1 and 8.2).

Nevertheless, as Pinch shows, public services remain a large and diverse sector in many local economies. In less-favoured areas public services ameliorate the effects of spatial centralization in the private sector. But an over-large non-market sector may also discourage private sector investment. National and local government expenditure cuts can also undermine over-dependent local economies. National government expenditure decisions are rarely consistent with urban and regional policies designed to support less-favoured areas. Thus more co-ordination between national and local economic policy over the geographical impacts of public sector change becomes increasingly necessary.

Further discussion and study

Guided reading

Jessop B, Kastendiek H, Neilsen K, Pedersen O (eds) 1991 *The politics of flexibility. Restructuring state and industry in Britain, Germany and Scandinavia.* Edward Elgar, Aldershot, Hants

Marshall J N, Alderman N 1991 Rolling back the frontiers of the state: civil service reorganisation and relocation in Britain. *Growth and Change* **22**: 51–74

Osborne D, Gaebler T 1992 *Reinventing government.* Addison-Wesley Reading, Mass

Pinch S 1985 *Cities and services: the geography of collective consumption.* Routledge and Kegan Paul, London

Activity

1. Information about the operations of public services is widely available, including how taxes or other levies are raised to support them, and the eligibility of various groups to receive them, by gender, income, age, medical condition, or area and length of residence.

    Collect such information locally, for one or more service. This might be for specific health treatment (e.g. in support for childbirth, care for the long-term sick); pre-school, school or higher education; entitlement to public (council) housing; rights of police protection and legal representation; land development control and environmental protection; refuse collection and disposal; state pension rights.

    (a) How is the service paid for; from national or local taxes?
    (b) What services are offered; who is eligible to receive them, and under what conditions?
    (c) Where are the services made available in different local areas?
    (d) What procedures are involved in applying for and receiving the service?
    (e) Are private sector commercial or professional agencies also involved, either in association with or as a substitute for public services? Are they controlled or regulated in any ways?

2. Review the local media for information on public perceptions of the quality of the services, and significant recent changes.
    Evaluate how far the following goals are satisfied in these cases:

    (a) equity of access
    (b) quality of service
    (c) efficiency of resource use

    Has this changed in recent years?

Consider

1. How do the goals of public sector agencies differ from private companies? Is 'privatization' of public services the best way of getting good value for money? Are alternative strategies possible?
2. Why has employment in public sector services become more significant in urban and regional development over the past 40 years?
3. There were significant changes in the character of public service *work* during the 1980s. Examine Chapter 8 and especially the evidence presented by Pinch and Winckler to identify the positive and negative aspects of these changes.
4. Pinch adapted Urry's (1987) analysis of modern restructuring processes in the private services to the public sector (see also Table 4.1). How effective is this, and how consistent are the processes he describes with those of:

   (a) industrial restructuring (Chapter 4);
   (b) the growing dependence on informal work (Chapter 5);
   (c) 'flexible specialization' (Chapter 6);
   (d) 'post-modernism' in consumption (Chapter 7)?

Discuss

1. The need to reduce taxes means that the subsidy from rich to poor and from core to peripheral areas embodied in all public services can no longer be sustained. What value judgements are implied by this statement?
2. If cities are the most efficient means of sustaining collective consumption, combining private and public agencies, what is their future in a period of declining public sector support?
3. Major public activities include branches of ministries and public agencies, local government, universities, hospitals and defence establishments. In spite of public sector rationalization, these remain a more stable basis for local economic development than almost any equivalent private investment. Discuss.
4. To what extent have studies of service location neglected public services?

Readings

The restructuring thesis and the study of public services

S P Pinch, 1989 (*Environmental Planning A*, 21: 905–26)

... What can be termed the restructuring thesis has come to dominate much of industrial geography (Massey, 1984), yet, until very recently, has been largely ignored in studies of public services and collective consumption. In this paper I argue that this state of affairs is undesirable. ...

An examination [is presented] of the relevance of the key concepts embodied within the notion of restructuring for an analysis of changes taking place in the public sector in contemporary Britain.

Restructuring and public services

Partial self-provisioning . . . The first category of restructuring is derived from the work of Gershuny and Miles (1983). These authors have drawn attention to the ways in which, as the cost of technology is reduced, certain services which were previously purchased outside the household can be replaced by the use of goods within the home. The classic examples are the substitution of videos and televisions for cinemas and 'live' sports, and microwave ovens for eating out in restaurants. The driving forces behind these changes are envisaged as essentially technological because these innovations affect the relative costs of goods and services . . . One can then conceptualise trends towards self-provisioning, not simply as the result of technological innovations, but as the result of efforts to move from collective towards individualised modes of consumption, reinforced by ideologies stressing 'individual responsibility' (Rose and Rose, 1982).

The extent to which communities and households are able to engage in self-provisioning has been the source of divergent trends in recent years. On the one hand, modern technology has produced many machines which ease the burden of domesticity, but the decline of the extended family and the breakup of traditional community networks have undermined the ability of people to cope with the burdens of caring for dependent groups. Whereas the key process behind increased self-provisioning in the realm of the private sector has been the desire for profit, the major imperative in the case of nonmarketed services in Britain has been the desire of Mrs Thatcher's Administration to reduce the size of the public sector by expenditure cuts and 'privatisation'. The extent of these processes would seem to vary considerably between different localities in Britain, especially as many Labour-controlled councils have resisted central demands for expenditure cuts (Pinch, 1987). The effect of these changes upon forms of self-provisioning is a key area for future research.

One new trend which might be encapsulated within the heading of self-provisioning is the increased reliance of local authorities upon unpaid volunteers to make up for staff reductions. These include volunteers who assist the social services and parental helpers in schools (Webster, 1985). So-called rationalisation of facilities (see below) has also increased pressures for self-provisioning. The closure of residential accommodation for the elderly, the reduction in forms of preschool provision, and the reduction of the school meals service have all put additional pressures upon families – and therefore usually upon women – to look after dependent groups. Last, one of the most important developments leading to increased self-provisioning amongst households is the development of so-called 'community care'. Although reducing the extent to which recipients of care are incarcerated in large institutions through the provision of smaller decentralised facilities in the community is in many respects a laudable aim, in practice 'community care' has often come to mean the closing down of older institutions with inadequate community services, thereby pushing the burden of care back upon families and informal networks within neighbourhoods.

Intensification . . . Intensification refers to increases in labour productivity via managerial and organisational changes with little or no investment or major loss of capacity. This concept is again highly relevant to the public sector where some of the most extreme examples of intensification can be found. In the British National Health Service (NHS), for example, efforts have been made to increase the rate of output for given levels of input. Thus, the number of patients treated per member of staff has been increasing. In the Wessex Region there has been a 9.3% increase in the number of patients treated, while the number of beds has fallen by 3%. This has been made possible by reductions in the length of time that patients spend in hospital and by a 25% increase in the number of day-care cases (Wessex Regional Health Authority, 1987). This higher throughput of patients has increased the workload upon the medical and nursing staff. In the case of nursing this has often involved a reduction in shift overlaps. Increased throughput is accompanied by a higher proportion of patients who are in the most dependent phase of their treatment, thereby increasing the demands they place upon staff. Increasing the numbers of dependent patients has put additional pressure upon junior doctors, making the long hours traditionally worked by this group even less desirable. In the Southampton and South West Hants Health District considerable efforts have been made to reduce the number of unsociable rotas, but, given that each junior post costs between £14 000 to £19 000 per annum, it has proved difficult to increase the number of posts within existing cash limits. Some nurses interviewed in Southampton claimed that it was the intensification of work rather than low pay which was primarily responsible for the low morale within the profession. As in many other areas of Britain, it has been difficult to obtain certain specialist and senior nurses. It has been suggested that one consequence of this intensification is absenteeism. In Southampton, absenteeism, defined as absence from work for reasons other than sickness, is not especially high but the absence of nurses because of sickness is said by managers to be 'causing concern.' Intensification has also affected clerical and administrative workers in the NHS. The increase in work loads has been associated with a reduction in the total numbers of administrative staff.

Intensification has also been under way in higher education, as the number of graduates per member of academic staff has been increasing in recent years (Urry, 1987). Contracting out of services, or at least the use of the threat of contracting out, has enabled many local authorities to intensify some of their spheres of operation, most notably in the context of direct works. In Southend, contracting out of refuse disposal enabled the abolition of the 'task and finish' system whereby workers completed the job after a 'task' rather than after a certain number of hours (Evans, 1985).

Intensification can also be brought about by a policy of redundancies, nonreplacement of retiring staff, or a greater reliance upon part-time work. All of these strategies put increased pressures upon a diminished work force. The pressures can be especially acute in a field such as social services where there are additional demands upon the remaining staff because of increasing numbers of the elderly, unemployed, and 'at risk' groups. Increased numbers of elderly and unemployed have also increased the demands upon the library service. Reductions in casual labour and overtime and increased staff redundancies have intensified the work loads of staff and have reduced staff cover for sick leave and holiday leave. In Bradford staffing in the library service is to be restructured so that less qualified junior staff have to take on additional responsibilities (Webster, 1985).

Investment and technical change Investment and technical change refer to situations in which heavy capital investment in new forms of production results in job loss. In comparison with manufacturing sectors, little is known about these forms of change in the private service sector, but many changes are underway in the sphere of nonmarketed services. These include the computerisation of health and welfare service records, the introduction of electronic diagnostic equipment in health care, and the possible use of distance learning systems using videos and computers in education (Gershuny and Miles, 1983). So far, most of this technology has had little impact upon the overall structure of employment in the public sector. New skills in the sphere of computers and software have been required and few jobs have been shed as a result (Gershuny and Miles, 1983). Many service jobs in the public sector, such as nursing and social work, involve caring for people and are therefore difficult to replace with technology. It was recently reported that in order to cope with shortages of nursing staff for intensive care, Guy's Hospital in London had started to use a computer-controlled system for altering drips. . . .

In the Southampton and South West Hants Health District the increasing emphasis upon new technology in medical care has led to demands for skilled operators and technicians. The NHS pays considerably less than the external labour market for such skills and there has been frequent 'poaching' of trained staff by the electronics industry which is an important element of the local economy. One response has been the progressive upgrading of jobs undertaken by workers after the introduction of new machines. This has the effect of raising salaries and helps the health authority to retain staff.

New technology has also been used in Southampton to reduce labour costs. This has primarily affected administrative workers through the introduction of computerised outpatient bookings systems, patient records, wages and salaries, and management information systems. This process will continue in the future and may extend to pathology where automation is seen to offer considerable potential for labour reduction. Part of this process of cost cutting has also involved the introduction of cook–chill catering systems. In most of these cases the NHS may be seen to be catching up with the private sector which has been much more ready in the past to introduce new technology to improve efficiency and reduce labour costs.

There seems to be general agreement that the jobs most vulnerable to technological change within the service sector are clerical jobs where word-processing systems can lead to staff reductions. Bradford Council is reported to have halved the number of typists employed, by the introduction of a centralised word-processing system (Webster, 1985). There are also fears about the impact of new technology in lessening job satisfaction through reduced promotion prospects and declining autonomy over working practices. It is interesting to note in this context that the introduction of new technology increased the problem of controlling the work force concerned with refuse disposal in Southend. Larger more efficient refuse disposal vehicles meant that workers could complete the tasks in a few hours and additional work had to be provided as overtime. The net result was that the workers had, by manual standards, relatively high incomes at additional cost to the authority (Evans, 1985).

Rationalisation Rationalisation refers to the closure of capacity, with little or no new investment. The most commonly cited examples of rationalisation in the context of private services are the closure of cinemas and laundries. There are, however, many more examples of rationalisation in the context of public services such as the recent closure of schools, hospitals, and welfare centres. Although many of the criteria involved are apparently rational and technical, as in the case of contracting out, closures are inherently political in character. Evidence from the USA indicates that, even when the criteria for school closure are based on technical issues applied in an explicit manner, the most detrimental effects are often upon the poorest communities (Honey and Sorenson, 1984).

There would seem to be wide variations in the extent to which it has been possible to cut back and 'rationalise' public services. Thus, there have been considerable reductions in nonstatutory services such as preschool child-care facilities and services for the elderly in some areas, but, as Mohan (1988a) notes, cutting back the NHS on any radical scale is not possible without damaging electoral consequences for the Conservative Government. Similarly, there are growing public pressures directed against the reduction of resources devoted towards education. Le Grand (1984) argues that the crucial factor affecting the vulnerability of a service for rationalisation would seem to be the class composition both of service consumers and of producers. When, as in the case of education and the National Health Service, there are both powerful middle-class producer groups and middle-class beneficiaries, the scope for rationalisation is limited. In the case of social work the beneficiaries are the poorer sections of society, but the providers are middle class, so these occupy an intermediate position in terms of their vulnerability to rationalisation. Those services such as refuse collection with working-class producers are easier to hive off to private contractors (see the next section).

In addition to class there is, however, another powerful explanatory factor which can be related to local government cuts – namely gender. Webster (1985) notes that not only are the services most vulnerable to cuts dominated by women providers but they are also the services in which the consumers who have most to gain are women. Women dominate many of the manual local government jobs that have been cut or contracted out and are left to shoulder the main responsibility as carers after the withdrawal of services, by acting as family carers.

In general, it would seem that there are a number of poorer, marginalised, and often discriminated-against groups that can suffer from public-sector policies of fiscal retrenchment. Thus, after the fiscal crisis in New York in the mid-1970s many of the blacks who had been recruited into the public sector through equal opportunity programmes in the 1960s were laid off (Sheftner, 1980).

Subcontracting ... Subcontracting is again a useful concept for analysing the public sector. Many of the private-sector producer services that have been growing so rapidly in recent years are, of course, a response to the privatisation (or more appropriately contracting out) of certain facilities such as cleaning and catering within the NHS. In Britain, subcontracting has usually led to a decreased work force, an intensification of the work process ..., inferior wage benefits (for example, reductions in sick pay and holidays), and the absence of trade unions (Spencer, 1984). It would appear that the main effect of the introduction of private contractors into the health service has not so much been a reduction in hourly wage rates as a reduction in the total number of hours worked, thus taking workers below the sixteen-hour legislation needed to qualify for employment-protection legislation (Mohan, 1988a).

The extent of privatisation has varied enormously between different local government areas, largely depending upon the political persuasion of the council (Moon and Parnell, 1986). In the Southend case the contracting out of refuse disposal was made possible by a combination of pressure from ratepayers and a right-wing council determined to gain increased control over the work force, albeit through the use of an external agency (Evans, 1985). In Dudley, the privatisation of the school meals service led to a reduction in the number of staff employed and a 40% cut in hours. Kent, Hertfordshire, Somerset, and Wirral have changed the conditions of service for part-time school meals staff by making them redundant and reemploying them to work shorter hours without the benefits of free lunches, holiday pay, and retainer payments during school holidays (Webster, 1985).

In the Southampton and South West Hants Health District the majority of catering, laundry, and domestic services put out to competitive tender have been awarded to the authority's own internal staff. Unions were hostile to the process of competitive tendering and adopted a policy of noncooperation. The process therefore involved local managers contacting workers directly to draw up detailed specifications for jobs which reduced costs to below those which could be offered by outside contractors. In Southampton General Hospital the number of hours worked by the cleaning staff have been cut by 25% with no reduction in the amount of work undertaken. The total number of cleaning staff was reduced from 337 to 252 and 43 workers on part-time contracts lost their jobs. The new job specifications have reduced restrictive practices and there is now more flexible deployment of staff. Part-time night staff who used to clean the casualty department have been dispensed with and emergency cover is now provided by cleaners who deal with operating theatres. Union officials argue that these new practices have led to a decline in staff morale and lower standards.

Paul (1984) claims that Medway District Health Authority deliberately declined to put ancillary services out to tender at hospitals where it was felt that union resistance would be particularly strong. In the case of Southampton General Hospital the ancillary services which have so far been affected by competitive tendering have all been dominated by female workers who might be expected to be less resistant to change than some of the male-dominated services that have had a history of 'difficult' industrial relations in the past. However, considerable changes have been made to working practices in these male-dominated ancillary services and it is intended to put security, portering, and driving out to competitive tender in the future.

The growth of part-time work ... The replacement of existing full-time labour by part-time work or by marginalised workers is a trend whose full significance in the context of public services has yet to be assessed. The rights of part-time workers are usually limited in relation to full-time workers, depending upon the number of hours that they work. In comparison with those working between sixteen and thirty hours per week, those working for less than sixteen hours per week lose rights of protection from unfair dismissal, entitlement to redundancy pay, and maternity provisions (Webster, 1985). Those working less than eight hours per week have almost no rights at all. It has been argued that the policy of reducing working hours to below sixteen hours has been most developed in the public sector (Labour Research Department, 1984). A number of local authorities have reduced the hours worked by part-time school meals staff to evade employment-protection requirements. There is also growing evidence that local authorities are making increased use of part-time teachers to overcome supply problems in particular areas. This would seem to arise because job losses in teaching have been concentrated in full-time jobs, whereas the number of part-time teachers has remained constant. Teachers who return to part-time work are generally restricted to the lowest grades on the Burnham scale and generally have fewer opportunities for training and promotion (Webster, 1985). The shift towards part-time employment seems to have been most pronounced in the libraries and museums service.

The significance of these developments is open to question, however, for different local authorities have responded in different ways to the same financial pressures (Travers, 1983). Some local authorities have concentrated staff reductions in the sphere of part-time work. These are predominantly, although not exclusively, the nonmetropolitan authorities such as East Sussex, and Hereford and Worcester. In marked contrast, other authorities – predominantly in metropolitan areas – have increased the number of part-time workers while reducing the number of full-time workers. Karan (1984) has argued that, in general, there is a shift towards part-time work because this gives local authorities greater flexibility to deal with financial uncertainty. There is also a pronounced shift towards the use of part-time workers within the NHS, although it is not altogether clear at present to what extent this reflects a response to financial stringency or to recruitment problems (Mohan, 1988b).

In the Southampton case the NHS makes far less use of part-time staff than the private sector does. In particular, private hospitals make extensive use of part-time nurses, whereas part-time nurses are almost entirely absent from the NHS. Although there are administrative problems in dealing with a large number of part-time workers in large units, the NHS seems likely to make much greater use of part-time workers in the future to deal with problems of staff shortages. An area in which there has been increasing use of part-time staff is administration. The NHS (together with other parts of the public sector such as the university and local government) pays below the going rate in the private sector for secretarial staff (a contrast of up to £2500 per annum at the top of the scale). The public sector thus has to make greater use of school-leavers and part-timers. In this instance the use of part-time staff is largely forced upon the health authority rather than being a chosen labour strategy. In some areas of ancillary services, contracting out has involved increased use of part-time staff. Interestingly, it has been found that increased use of part-time staff in this way can lead to less flexibility in terms of labour deployment. Whereas in the past, low-paid full-time workers had been prepared to work overtime to supplement their wages, it has proved more difficult to persuade many part-time workers to adopt more flexible working rotas. It is assumed that this rigidity arises because of domestic constraints upon such workers. It is interesting to note in this context that part-time cleaners taken on by private hospitals are expected to be on call when needed at almost any time and there is no shortage of women prepared to undertake such work in the Southampton region.

The enhancement of quality and the materialisation of service functions Question marks also hang over the significance in the context of the public sector of . . . the enhancement of the quality of services and the materialisation of service functions. Certainly, efforts have been made in certain spheres of the public sector to increase the quality of services through retraining schemes. Some of these schemes have included instruction of personnel in issues related to race and gender and their implications for local authority service delivery. Efforts have also been made to make local authorities more responsive to the needs of communities through neighbourhood participation schemes and the establishment of decentralised neighbourhood offices. Much of the initiative for these schemes has come from councils that have been dubbed 'The new Urban left' (Gyford, 1983) and these changes stand in stark contrast to the general push towards cost reduction. The effect of privatisation and subcontracting upon the quality and efficiency of public services is of course an enormously controversial issue at the present time. Two recent reports indicate that privatisation of telecommunications (British Telecom Unions Committee, 1986) and water services (Patterson, 1987) has led to a deterioration on the quality of services for private consumers compared with industrial interests.

Spatial relocation, centralisation, and domestication In his recent paper Urry (1987) has extended his early typology (Urry, 1986) with three more processes at work in the context of services. Two of these, spatial relocation and centralisation, are closely related processes which are long established in the context of public services. . . . Domestication is another concept which is closely related to the partial self-provisioning described above. Clearly, many of these processes . . . are closely linked and will be taking place simultaneously in any particular service. . . .

Conclusions

In this paper I have highlighted the ways in which the extensive battery of concepts that have emerged from the study of industrial restructuring can be used to analyse contemporary changes in the structure of welfare systems. The major advantage in linking concepts from industrial restructuring to the analysis of changes in the public sector is that they reveal how diverse are the changes in the nature of the welfare state in recent years. . . .

The major limitation of ideas from industrial restructuring in the context of public services is that they focus upon outcomes rather than on pressures for change. Although many of the outcomes in the public sector parallel those in the private sector, the pressures upon public-sector organisations are very different. Furthermore, the range of the types of change taking place in the public sector would seem to be less than in the private sector. Because of the political and fiscal constraints upon local authorities they have less capacity for closing 'plant', shifting product lines, making redundancies, or investing in new equipment. . . .

Restructuring the civil service: reorganization and relocation 1962–1985

V Winckler, 1990 (*International Journal of Urban and Regional Research* 14: 135–57)

. . . Little attention has been paid to restructuring in the various public sector services, even though they have been the subject of repeated attempts at reorganization, both *in situ* and spatially. This is regrettable. . . . The public sector warrants analysis in empirical terms, not least because of the sheer numbers of the UK workforce it employs. In 1981 over 35 per cent of employment in the service sector in the UK was in the four public sector-dominated industries of medical and health services, education and national and local government service. . . .

The public sector is of course very varied. In the UK it includes a range of services from health and education, to fire fighting and public sewage, to the administration of local and central government. The last of these, namely the UK civil service, is the focus of this paper.

Sectoral context: the UK civil service and the state

The UK civil service has a specific role within the public sector, which is to act as the nonpartisan administration of political decisions. This role has not only structured the characteristics of the civil service itself, but has also profoundly shaped the course of its restructuring, including its spatial reorganization.

The civil service is the apparatus through which the state administers its policies, and is thus involved in the direct implementation of some policies (for example the payment of social security benefits) as well as in overseeing the implementation of others, such as regulating activities carried out by nonstate bodies. . . . The government of the day can therefore exert direct control over the civil service through its position as manager and via the hierarchy of accountability, and can hence determine the nature of any restructuring. . . .

The civil service's role means that both the overall size of the task it faces, and the precise areas and means of implementation, are determined by government policies and priorities. Issues of need or profit do not enter directly into the organization of civil service work. Shifts in the level and nature of government intervention therefore have an immediate impact upon the tasks the civil service has to perform and, thus, on the size of the civil service.... Significantly, the state itself is the employer.... This gives the state a direct input into conditions of work. Indeed, civil servants are not strictly employees at all (they do not have formal contracts of employment) but are 'servants of the crown' (Corby, 1984) with a duty of 'undivided allegiance to the state (CMNI) 7057, 1978: 3). The term civil servant covers a vast range of occupations, from scientific posts to professional occupations such as lawyers and accountants, through all levels of administrative posts to a substantial manual workforce such as porters and cleaners. By no means all civil servants enjoy good pay and conditions for their labours: indeed it has been estimated that one in five nonindustrial civil servants works for low wages (defined as two-thirds of median earnings of all males).

... The central pressure on the civil service is experienced by most nonprofit state services, namely there are demands to socialize and therefore expand certain activities at the same time as demands to minimize their cost (Gough, 1979). ...

The contradiction between expanding activities and constraining their costs has been constant throughout the postwar period. What has varied, and is of particular interest here, is the precise *form* that the contradiction has taken and the specificity of various responses to it. In each case, the form of the central pressure on the civil service, and the responses to it, have been conditioned by the role of the civil service itself. Two distinctive periods can be identified: the first running from the early 1960s to the mid 1970s, the second from the late 1970s to date. The former is marked by increased state intervention, a concomitant expansion of the civil service and the use of strategies which permitted expansion within certain limits. I have termed these strategies 'accommodating strategies'. The latter period has been rather different, with a general reduction in state activity, a contraction of the civil service and the adoption of strategies to restructure the civil service which have been geared to a more fundamental reorganization. I have termed these 'transformative strategies'. Of particular interest is the marked contrast in the extent to which spatial strategies have been utilized as part of the different forms of restructuring.

Expansion and accommodation: the growth of the civil service

It is of course widely known that state intervention in many spheres increased from the early 1960s to the mid-1970s. Postwar demographic shifts brought an increase in the child and elderly populations, both heavy users of the major welfare services which expanded accordingly (Gough, 1979). At the same time state provision extended into new areas, from the expansion of higher education to the introduction of redundancy payments. State intervention in the economy also increased during this time (Budd, 1978; Gamble, 1981), through financial aid to industry, for example through regional policy, and through direct servicing, such as the establishment of a Business Statistics Service in 1966 (Berman, 1981; Huckfield, 1978).

What is less widely recognized is that in each case the civil service planned, administered and sometimes implemented the new or expanded service. By 1975 the civil service was responsible for handling services involving £26 417 million of state expenditure (Gough, 1979: 84). These increases in state activity were reflected in a substantial increase in the size of the civil service. Between

Table 1 The size of the civil service 1965–84

	Number of civil servants	Percentage change
1965	667 980	–
1970	702 056	+5.1
1974	694 384	–1.1
1977	746 161	+7.5
1979	733 176	–1.7
1984	632 591	–13.7

Note: Groups which have ceased to be defined as civil servants over the period have been excluded throughout as far as possible.
Source: Civil Service Department, 1975: Table 2; HM Treasury, 1984a: Table 2.

1965 and 1977 the civil service expanded by 78 000 staff (see Table 1), to reach a peak of almost three-quarters of a million people.

The expansion in the civil service over this period was not, however, entirely unchecked. . . . Strategies were developed which aimed to *contain* the growth of the civil service. By this I mean that they attempted to minimize its expansion without fundamentally challenging its activities or organization: I have therefore termed them 'accommodating strategies'.

Internal reorganization The long-established tool of reviewing and reorganizing the civil service was used frequently and to considerable effect in the period of expansion. The departmental structure of the civil service was scrutinized and subsequently changed in 1966 and again in 1970. By far the most thorough review of the civil service in the postwar period was that set up in 1966, under Lord Fulton. Its remit was to, 'ensure that the [civil] service [is] properly equipped for its role in the modern state' (Fulton Report, 1968: 107). It advocated, in keeping with the general trend of the times, greater centralization of management, the use of modern management techniques and greater staff specialization. Four years later another review, this time published as a White Paper, was, in contrast, more concerned with greater policy efficiency. Here too there was an emphasis on centralization, one result of which was the merger of several government departments into the massive Department of the Environment (CMND 4506, 1970). . . .

The office labour process A second set of strategies were those concerned with changing the nature of the office labour process in the civil service. Although superficially quite dramatic in impact, their main achievement was to allow the continued evolution of the civil service. The techniques used to change the office labour process parallel developments in manufacturing, notably the development of a fine division of labour and large-batch production. None of these techniques was new: fragmentation of the labour process into smaller and simpler parts has long been a feature of civil service work organization, not least because it eases the application of bureaucratic procedures. Indeed, the Fulton Committee noted the widespread division of labour into 'repetitive elements of the smallest possible content', and suggested that it did not necessarily make the best use of labour resources (Fulton Committee, 1968: 38). Rather than transforming the labour process, such techniques simply eased management and control.

A somewhat newer departure was the large-scale introduction of computers to the civil service in the 1960s. The contemporary generation of computers was particularly well suited to the processing of large amounts of simple data, and it was in this area that the first computer applications were made, with considerable enthusiasm. Civil service investment in computers doubled every two to three years in the 1960s (Civil Service Department, 1971a) so that it was, and remains, one of the largest computer users in the UK (Owen, 1974; Atkinson, 1980). In many ways, computerization was an accommodative strategy *par excellence*. Its main attraction was not to reduce staffing – indeed early applications tended to generate large numbers of routine backup posts sometimes exacerbating existing recruitment difficulties (Civil Service Department, 1971a) – but was rather its ability to carry out a much larger volume of work. . . .

Spatial reorganization   The third accommodating strategy – the relocation of the civil service – was perhaps the most radical of the period. Although some offices had been dispersed away from London for strategic reasons during the second world war, most were re-established in London in the 1950s (Hammond, 1967). By the early 1960s the civil service was highly concentrated in the London area. . . .

. . . Plans for a new phase of civil service dispersal were announced in 1963 (House of Commons Debates, 18 July 1963: Volume 681, Written Answers Columns 82–86). Altogether some 64 000 jobs were relocated or newly established away from London between 1963 and 1979, two-thirds in the assisted areas. Most of these were large blocks of clerical work (House of Commons Committee on Scottish Affairs, 1980; Hardman Report, 1973). Whilst this undoubtedly sounds like an impressive achievement, a measure of the extent to which the strategy coped with *excess* civil service numbers rather than effecting a fundamental spatial reorganization is that in 1977, after 12 years of dispersal policy, the southeast of England had a *higher* concentration of civil service posts (41% of the total) than in 1965, and virtually twice as many staff as in 1965 (Civil Service Department, 1977). It seems that far from bringing about a redistribution of civil service work, dispersal policy simply accommodated the continued expansion of lower-grade work and demands for regional development.

Dispersal was undoubtedly a response to pressures on recruitment in the civil service in London. Various attempts to attract staff, for example by lowering entry qualifications, raising the age limit for entry to clerical posts, and extensive advertising, failed to produce the required number of applicants (Civil Service Commissioners, 1962; Civil Service Commission, 1966). In contrast it was noted that there were plentiful well-qualified applicants elsewhere, and suggested that the long-term solution to recruitment difficulties lay in reducing labour demand in London by moving clerical work (Civil Service Commissioners, 1962: 3; Civil Service Commission, 1971: 10).

The availability of female labour in peripheral regions was a central consideration to the relocation strategy, although, as has been pointed out (Bowlby *et al.*, 1986; Sayer, 1985), the relationships between female labour supply and locational change are very complex. Certainly in the civil service there is no evidence of a preference for or discrimination in favour of women staff. On the contrary, the concern to recruit in an unbiased way the nonpartisan staff the civil service needs to fulfil its tasks, has resulted in the basic principle that each job application is treated entirely on its merits, backed up by elaborate procedures. There is no formal preference for males or females (Civil Service Department, 1971b).

Nevertheless, there are marked gender divisions *within* the civil service, and it is on these (and the divisions in the external market) that the civil service attempted to solve its recruitment crisis through dispersal. The recruitment crisis was essentially a crisis in the female labour market. The main staff shortages were in the clerical and secretarial grades. Women constituted three-quarters of the entrants to the clerical assistant grade in 1975 and 80 per cent of that grade were women (Joint Review Group on Employment Opportunities for Women in the Civil Service, 1982). The civil service was well aware that clerical work was a female-dominated occupation, for example aiming some of its recruitment at married women returners. Significantly, the clerical and secretarial grades also formed the bulk of dispersed work. The civil service was well aware that female labour was more widely available outside the southeast of England. An appraisal of the relocation programme in the early 1970s selected locations for dispersed civil service offices according to their 'capacity' to recruit and retain clerical staff: crucially, this was measured as the amount of total and female office employment already present, on the grounds that most dispersed jobs would be filled by women, as indeed they were (Hardman Report, 1973; Hammond, 1968; Winckler, 1986; Civil Service Department, 1977). The civil service thus incorporated existing spatial and gender divisions into its restructuring strategies whilst itself maintaining its gender neutrality.

The dispersal process is also more complex in other ways than has been suggested. Far from relocating to seek 'cheap labour', the nationally determined pay and conditions in dispersed offices were often superior to those of other types of *women's* work in receiving locations (Hammond, 1968; Winckler, 1986). Recruitment to posts in dispersed offices was easier, staff were better-qualified, staff turnover was much lower and the quality of work much higher (Civil Service Commission, 1971; Civil Service Department, 1977; House of Commons Committee of Public Accounts, 1981: 6). Dispersal also offered an opportunity to reorganize working practices. Clerical and related staff (usually married women) used to well established work methods rarely moved with the office and were usually transferred elsewhere while new workers could be taken on in the new location and trained in new methods (Hammond, 1968).

Lastly, dispersal was frequently carried out in conjunction with computerization. In addition to all the advantages of reorganizing working practices mentioned above, dispersal eased recruitment to the large numbers of clerical posts computerization tended to create (Civil Service Department, 1971a: 13). Conversely, technological change also facilitated dispersal by allowing functions to be split between a London-based head office and regional backup facilities, as in the case of the Companies Registration Office (Allwood, 1980: 78).

Of course, the spatial reorganization of the civil service was not simply a response to regional variations in labour supply and the benefits of relocation. The political level, to which the civil service is peculiarly open, also had a major input into the formulation and implementation of this strategy. The input was the regional dimension.

Dispersing the civil service was both a relatively easy way of creating jobs in the regions and of heading off regionalist demands. Labour administrations in the period claimed that the assisted areas had priority for dispersed work (e.g., House of Commons Debates, 4 June 1965: volume 715, Columns 2176–92; 30 July 1974: Volume 878, Columns 482–94), with 90 per cent of dispersals following the Hardman Report being destined for assisted areas. . . .

In contrast, Conservative administrations at the policy's inception and in 1970–74 emphasized the importance of proximity to London in deciding office locations (e.g., House of Commons Debates, 18 July 1963: Volume 681, Written Answers Columns 82–86; 19 December 1971: Volume 827, columns 1497–98). . . .

Contraction and transformation: the civil service in the 1980s

The latter part of the 1970s saw a reorientation of state policies on the economy in general, state intervention and the civil service. The growing fiscal crisis of the state (O'Connor, 1973) finally brought a response in the form of attempts to limit public expenditure from 1975 onwards. The return of the Conservative administration in 1979 saw a more explicit and ideological commitment to minimizing state intervention and reducing public expenditure. The public sector as a whole was cast as an 'unproductive' burden on the economy (Bacon and Ellis, 1978) to be returned to the market or the family, or made more efficient. Regional development was no longer a political priority, and regional aid was steadily cut.

The pressures on the civil service under these conditions were even greater than before, although they took a different form. This time the pressure was experienced directly in financial terms, rather than as a lack of competitiveness in the labour market. The civil service was close to its peak in terms of numbers, and costs were mounting. The nature of its work meant that it appeared to be particularly unproductive and inefficient, and it enjoyed little public or political support. As before, the civil service was singled out for restructuring, again not least because of its openness to political intervention. Quite simply, the civil service was seen to be an exemplar of the inefficient, unproductive and costly public sector. The government's objectives for the civil service were summarized as follows:

> The Government took office determined to improve the efficiency of the Civil Service, to eliminate waste and to promote the methods of administration which enable and encourage staff to give the best possible value to tax payers. (Privy Council, 1981: 1)

The concern of the government over the last 10 years has been to reduce the size and therefore the cost of the civil service in a variety of ways. But, whereas previous strategies 'accommodated' changes in the size of the civil service, here the objective of strategies has been to *transform* the civil service into a small, highly efficient and cost-effective administration. The forms that these strategies have taken have been shaped by the role and characteristics of the civil service.

Manpower control One of the crudest yet most effective techniques used in pursuit of the government's objectives has been that of 'manpower control'. Indeed, it hardly warrants the status of a restructuring strategy, for the reduction of the size of the civil service was an end in itself, to be achieved by any means. After the 1979 election there was an immediate ban on recruitment to the civil service and a reduction in its size of 14% (102 000 jobs) over five years was announced. This was followed by a further 6% reduction 1984–88 (HM Treasury, 1984b). The relationship between manpower control and the wider goals of the government is clear in this extract from Civil Service evidence to the Select Committee of Public Accounts in 1984:

> the manpower targets were an expression of the Government's priorities and the Government had as one of its priorities the reduction of expenditure ... the decisions on manpower and other resources were bound to be taken in the light of that view on priorities ... [W]ithout those targets probably not the same reduction would have been achieved. (House of Commons Committee of Public Accounts, 1984: 20)

This strategy was only possible because the government had direct and immediate control over the civil service *as an employer*. So strong was the government's commitment to the reduction in size (and perhaps its conviction that there was spare capacity in the service) that this strategy was implemented without regard to the nature of the tasks which the government was setting for the civil service, in complete contrast to earlier years.

Reorganization The reorganization of the civil service has continued but in a quite different way. The emphasis has been on 'distancing' administrative functions from the state. Again, full use has been made of the government's position in control of the tasks of the civil service. Most drastic of all has been the complete elimination of some activities. Various reviews of 'the need for work' in the civil service to 'eliminate those functions that are unnecessary or are no longer required' (CMND 8616, 1982: 1) have been carried out. One review of civil service work is estimated to have made a one-off saving of £20 million and a further saving of £130 million annually.

Some functions have been moved out of the civil service into other parts of the public sector or into the private sector. Even small-scale moves of individual tasks have saved posts: for example, shifting the administration of housing benefit to local authorities and the responsibility for sickness benefits onto employers 'saved' 6000 civil service jobs (House of Commons Committee of Public Accounts, 1984). 'Hiving off' whole sections of civil service work into the public sector has also 'saved' civil service jobs. 'Privatization' is a strategy which has received a good deal of attention, not least because it represents an about turn in the whole ethos of formerly civil service activities. However, at least until recently, it had actually made very little impact on the civil service. Despite the rhetoric, hiving off and privatization were hardly spectacular in effect, accounting for an estimated 12% of total job loss (HM Treasury, 1984b). The impact of proposals made in February 1988 to hive off large parts of the civil service into self-funding and managing agencies will have a much greater impact, although at the time of writing it is unclear to what extent the plans will be implemented. . . .

Management practices and the labour process Similarly, management practices in the civil service have received renewed attention. Management is seen to be particularly ineffective in the civil service because of its lack of profit motivation and bureaucratic principles. The government's aim has been to make the civil service 'efficient and effective', the former being defined as, 'the maximization of output in relation to input' and the latter as, 'the extent to which activities achieve their objectives' (House of Commons Treasury and Civil Service Committee, 1981: 38). . . .

Underlying several innovations in management practice is a fundamental shift in management principles, involving a move away from bureaucratic rules towards management according to financial demands. Financial management has become more sophisticated in the 1980s. Early centrally administered cash limits on operations have been refined to a set of locally implemented management techniques which combine highly centralized control over total expenditure (marked by the absorption of civil service central management into the Treasury in 1981) with devolved responsibility and assessment of performance (CMND 9058, 1983). Such local management and the principles on which it is based are unprecedented in the civil service.

Changing management techniques have also contributed towards reducing the size of the civil service: over half the reduction in the first phase has been attributed to changes in working practices (HM Treasury, 1984a: p. 7). Some relatively minor changes have resulted in large financial savings, for example the streamlining of procedures of simplification of forms. Eliminating one form relating to redundancy payments is estimated to have saved £250 000 (CMND 8504, 1982). More recently there have been proposals to radically reform civil service working patterns, with a view to increased flexibility of labour (Brindle, 1987).

Computerization is playing a major part in restructuring: but in the 1980s the objective is to reduce costs through cuts in staffing rather than improving the quality or quantity of service. All computerization projects are assessed for their 'profitability' before they are approved, with a particular emphasis on 'quick returns' (House of Commons Treasury and Civil Service Committee Sub-Committee, 1981: 177). Computerization in some form is underway in almost all departments, the largest being in the Inland Revenue (HM Treasury, 1984c) and the Department of Health and Social Security, each of which is estimated to 'save' 6000 posts (House of Commons Committee of Public Accounts, 1984). . . .

Relocation The dispersal of the civil service is a particularly interesting strategy over this period. It was one of the first casualties of limitations on public expenditure and subsequent cuts in the size of the civil service. The 1974 programme of dispersal had already been rephased in 1977 because of spending cuts (House of Commons Committee on Scottish Affairs, 1980: Annex 2) and in 1979 it was abandoned altogether. The reasoning was that the programme was so far removed from the principles of the Hardman Report, which gave priority to efficiency considerations, that it was 'impossible to see the justification for it' (House of Commons Debates, 26 July 1979: Volume 971, Columns 902–22). Clearly, plans to cut the size of the civil service meant that labour demand would also be substantially reduced whilst rising unemployment had, even by the early 1980s, eradicated recruitment problems in London and the southeast of England (Civil Service Commission, 1979; 1980). Regionalist demands had little purchase on a government with relatively small representation in Wales, Scotland and the north of England, and which was committed to reducing regional aid. Moreover, soaring unemployment and a weakening of trade union power, especially in the civil service itself after the unions' defeat in the 1981 pay dispute, meant that there was limited opposition to plans to introduce new management techniques and working practices *in situ*. What opposition there has been has largely failed. In other words, when the political priorities, internal pressures in the civil service and patterns of uneven development which had formerly favoured a spatial restructuring strategy evaporated, the strategy itself expired.

However, there are recent indications of a return to a relocation strategy, albeit in a different guise and by no means as a coherent policy. There are reports of acute staffing shortages in the clerical grades, high staff turnover and low morale in London offices. At least one office, the Patents Office, is to move from London to Cardiff, south Wales in one of the first dispersals for almost 10 years. Further relocations are said to be possible, although unlike earlier moves, they will be considered at *departmental* rather than governmental level. This change of course fits in with the general decentralization of management, and suggests that regionalist political pressures, in so far as there is any susceptibility to them, take a different form. The recent revival of a spatial element in restructuring appears to confirm that it is very much a response to uneven development, as indeed does the cessation of the policy in the early 1980s.

So, whilst the nonspatial strategies used to restructure the civil service –

namely manpower control, reorganization, changing management practices and work organization – are broadly the same *type* of strategies as in the earlier period, they are quite different in both their *content* and *objectives*. The over-riding concern has been to reduce the size of the civil service and transform its character. Much lower priority has been given to spatial reorganization, because there appeared to be neither the political nor economic pressures to consider it. . . .

Conclusion

It is clear that there are strong pressures to reorganize in the service sector and in the public sector, and that those pressures have been exerted even in a part of the public sector which is relatively 'immune' because of its privileged place. However, the form that restructuring has taken has been conditioned by the role of the civil service and its organizational characteristics. Economic pressure has been experienced primarily in fiscal terms, and in particular in terms of the cost of the civil service. At certain times, this has been translated into recruitment problems as civil service pay failed to compete in tight labour markets. Similarly, the form that restructuring has taken has been shaped by the civil service's role and characteristics. Most changes have been organizational, be they changes in the structure, operating principles or spatial distribution of the civil service. State policy towards the public sector in general has been crucial here, with the civil service's role as the state administration making it especially open to shifts in policy.

It is also significant that uneven regional development has played a part in the formulation of a relocation strategy in the public and service sectors, as well as in the more widely studied case of certain manufacturing industries. Recruitment difficulties, high office costs and regionalist lobbies were the ways in which uneven development impinged on the civil service.

CHAPTER 9

Services as wealth creators

This final chapter pulls together the implications for wealth creation, at various geographical scales, of the aspects of service activity reviewed in Chapters 5–8. Modern economic restructuring has radically transformed the prospects of localities, regions and even nations over the past two decades. In our view, service functions play an important, and sometimes leading, role in these changes, although the key types of interrelationship through which they influence events are different at various scales. Our primary interest is in comparative patterns of wealth creation in urban or regional economies, but these obviously cannot be isolated from wider national and global developments. Here, the wealth-creating roles of services will be reviewed at different geographical scales, while also emphasizing the important interdependencies between them and manufacturing.

A review of the argument

From our perspective the traditional sector view of service activities is inadequate since it consigns them largely to the support of consumption, rather than the wealth creation associated with primary and manufacturing production. Even in the eighteenth century, this did scant justice to the productive role of service skills, and it soon became recognized that transportation and trading expertise were also important for the development of production (Chapter 3). Nevertheless, the separation of services from wealth creation persists even today in much economic and political debate (Fothergill and Gudgin, 1982; Peck and Tickell, 1991).

Nowadays, as we have seen, service support to material production has been extended to include the provision of technical, including communications and computer expertise, as well as the whole variety of management functions (Britton, 1990; Daniels and Moulaert, 1991). Such 'business services', with the parallel functions provided for their own use by manufacturers, are particularly important in innovative and successful sectors such as energy, chemicals, pharmaceuticals, electronics, aerospace and motor vehicles (Marshall *et al.*, 1988).

Other types of service also play their part. Even the quality of apparently consumer-oriented activities, such as marketing and retailing, affects manufacturing success, especially in wealthy societies, where competition is most intense (see Chapter 7). Public sector services, such as education and training, or health care, also sustain the knowledge, skills and quality of the workforce (Chapter 8). Many types of service are thus integral to the operations and competitiveness of the more obviously wealth-creating extractive and manufacturing activities.

Do services, then, only support other more 'basic' functions, rather than being a source of wealth themselves? We saw in Chapter 4 that Walker at first seemed to argue this, with services creating surplus value only to the extent that they play a role as 'indirect labour' for material production (Walker, 1985). However, he also described this realm of indirect labour as possibly the 'principal locus of industrialism'. In his view, although service expertise must serve some ultimate materially related outcome to create wealth, such expertise has become increasingly significant for the success of production. We take this argument further, and see the economic contribution of services as determined by a balance of mutual dependence between different manufacturing and service functions, within which, in many circumstances, the latter may well take the lead. This is often the case for business and financial services, but even trends in consumer and public sector demand (e.g. in tourism, health care or information services) now stimulate manufacturing innovation. Thus, service functions are of growing significance, both quantitatively, in employment and output terms, and as a significant element in economic change. However, the deep linkages between services and other activities means that the notion of a separate 'service sector' has no coherent justification.

The significance of service–client interaction varies for different types of activity under different organizational arrangements, in different places, and at different times. Hence while the expertise offered by a service may sometimes merely support wealth-creating innovations originating elsewhere, at other times it may supply the vital initiative. The weight of responsibility depends on the particular social and economic context of the many functions which are combined in production. For example, the high-level skills required for managing strategic change at a company's headquarters are generally more critical for wealth creation than routine distributive services at the local supermarket. Some regions specialize in such higher-level managerial and technical functions, without any obvious direct local connection to material production, and their presence may even influence manufacturing to invest there. The growing importance of service trade over longer distances, means that in some circumstances they may now offer sources of wealth that are as self-generating as any comparable extractive or manufacturing activity. This has for long been true of the entrepôt functions of major ports, and is increasingly the case for major financial centres, such as New York, London and Tokyo. Other service functions, including tourism, business services, government functions or specialized consumer services may play a similar wealth-creating role.

'Manufacturing matters' versus the 'new service economy'

Discussion of the significance of service-based wealth creation has been strongly coloured by the active debate since the 1970s over the decline of manufacturing in the US and Europe (Petit, 1986; Harris, 1987; Cohen and Zysman, 1987; Malecki, 1991; Patton and Markusen, 1991; Fik *et al.*, 1993). Many analysts, from industrial lobbyists to both conservative and Marxist commentators, have argued that successful economies must retain significant extractive or manufacturing capacity. Future wealth depends primarily on responding to competition from Japan and the newly industrialized countries by restructuring manufacturing towards high-technology and other high value-added products.

'Manufacturing matters' was the clarion call of Cohen and Zysman's (1987) much-discussed book on the subject. By extension, although with rather less logical basis, regional development policies have also been primarily oriented to renewing or even creating a manufacturing base. Old industrial regions have declined, and 'new industrial spaces' emerged, as models for others to follow, specializing in manufacturing activities more effectively adapted to modern global competition (Scott, 1988a; Lovering, 1990; Benko and Dunford, 1991).

Others have argued that the 'new service economy' is itself a consequence of adaptation to changing global competition (Shelp *et al.*, 1984; Riddle, 1986). Know-how and expertise are the greatest assets of the established advanced economies, and are increasingly traded world-wide, as well as being vital to the competitiveness of old and new industries at home. Financial expertise is particularly important in the capitalist-dominated global economic system, and has enabled a few countries, and regions within them, to control the recycling of global capital resources. Some of the most successful regional economies in the 1970s and 1980s, were those, mainly in and around major metropolitan centres, which developed a global financial and business service base (Daniels, 1986b; Gillespie and Green, 1987; Drennan, 1989; Harvey, 1989a; Warf, 1991; Wood, 1991b; Noyelle and Pearce, 1991). They exploited the triple agglomeration advantages of (a) information exchange and processing, (b) skilled labour training and deployment, and (c) investment in transportation, data processing and communications infrastructure (Stanback *et al.*, 1981; Stanback and Noyelle, 1982). As much as manufacturing, these areas draw wealth from national and international markets, by organizing, renewing and marketing scarce expertise.

Both the 'Manufacturing matters' and the 'new service economy' scenarios contain some truth. The terms of the debate, however, still

215

echo the old 'sector' perspective, failing to acknowledge either the complexity of modern manufacturing, or the diverse roles performed by services in the contemporary economy. We have seen that modern manufacturing depends on a variety of service functions for success, whose wealth-creating role is often embodied within nominally manufacturing products, as well as other services. At the same time, some information-based services are directly tradable, and therefore may be sources of growth for some regions, and even nations. However, the information they handle has no economic value until it is applied elsewhere, for example to help design or market a product, and these services depend on material innovation, for example in communications and computer technology, for their success.

To sum up, wealth in the modern global economy depends on mustering and marketing a diverse range of technical and organizational service skills and applying them effectively to valued end-uses, including a range of innovative manufacturing and tradable service functions.

Service exchange and wealth creation

Wealth creation within any area, whether it be a nation, a region within a nation, or a locality within any region, depends on earnings gained from exchange with other areas. At each scale, the 'economic base' which determines an area's wealth, consists of those goods, service skills or capital resources that it can produce to sell elsewhere, whether in the next town, other regions of the same country or abroad.

Exchange may arise from various types of transaction, including;

1. direct trade, under which goods or services made in one place are sold elsewhere;
2. the returned earnings of migrant labour, and local expenditure by other mobile groups such as tourists;
3. the returned interest, dividends or profits of capital investment taking place elsewhere.

The relative contributions of direct trade, human mobility and investment may differ according to the spatial scale under consideration, but the same processes of wealth exchange apply. At the international level, exchange is dominated by trade in goods, although 'invisible' earnings, especially the exchanges of investment income, have expanded rapidly through the spread of multinational and global corporations (Box 9.1). Service exchange generally takes more diverse forms than goods. This makes it more difficult to measure.

Larger nations or regions are likely to be more self-sufficient than smaller areas. In general, more exchange takes place over shorter distances, so that interlocality and interregional transfers represent a higher proportion of each area's output than international trade. While service exchange takes place at all scales, its significance even at the

---

Box 9.1: The international operation of services

It is difficult to measure the increasingly international operation of services (Daniels, 1991b; 360–1; 1993: 77–81). Nevertheless Daniels (1993) documents the growing involvement of services in international trade, and shows that almost 81 per cent of the total world exports of services originate from just 20 countries. The same countries, although in a different rank order, are amongst the top 20 importers of services (77 per cent of total world imports). Though there is considerable variability in the flows of service trade from year to year, the US dominates global service transactions: more than one in ten transactions (by value) either originate there or are destined for US customers. France, the UK, Germany and Japan are other significant exporters and importers (Daniels, 1993: 82).

The contribution of goods to world current account transactions declined from 73 to 68 per cent between 1969 and 1986, indicating that the contribution of services to trade has increased. More recent estimates from GATT suggest service exports increased from 560 billion dollars in 1988 to in excess of 600 billion in 1989. There appears to be a relationship between the level of economic development of a country and the propensity to export services and to generate a demand for service imports. But the growth in services trade is concentrated in a narrow range of services. In the 1980s, only business services, financial intermediation and shipping and transport have shown a relative increase in their share of world trade; the share of trade in other services has remained static.

Trade data underestimate international service transactions, many of which take place within companies. Foreign direct investment (FDI) statistics (that is, the investment abroad by multinational companies) give some idea of the scale of such transactions. It is difficult to separate the services element of FDI, but it is likely that the 'similarities [between services and manufacturing] are more striking than the differences' (Sauvant and Zimney, 1989; quoted in Daniels, 1993: 90). FDI increased by twice the rate of gross national product (GNP) in the OECD countries in the 1960s and by four times the rate of GNP between 1986 and 1988. By 1995 it is estimated that the annual flow of FDI will amount to $228.8 billion (Julius, 1990).

The growth of FDI is taking place largely between developed countries, the same countries prominently involved in international trade. Some 75 per cent of the world stock of FDI is held by the USA, the UK, Japan, Germany and France. Over time the importance of the USA and the UK in FDI is declining, and Japan is becoming increasingly important as a source of outflows. It is believed that services form an increasing proportion of FDI, with finance, telecommunications and business services especially prominent (Daniels, 1993). Taking the evidence of trade and FDI together it is clear that services have participated in a significant shift towards a new level of integration in the world economy (see also Clairmonte and Cavanagh, 1984; Dicken, 1992: 349–82).

---

international scale was only fully appreciated in the 1980s, when service trade became a major issue in international trade talks (Nusbaumer, 1987; Office of Technology Assessment, 1987; Bhagwati 1987; Tucker and Sundberg, 1988; Feketekuty, 1988; Nicolaides, 1989; Daniels, 1993). International exchange is also an increasingly important component of the wealth-creation capacity of services in regional and local economies (Keil and Mack, 1986; O'Connor, 1987; Goldberg et al., 1989, 1991; Perry, 1991; Porterfield and Pulver, 1991; Begg, 1993). The growth of international service trade, as we shall see, has created new types of service-based region (Allen, 1992; Drennan, 1992), but any regional economy nowadays may include some sort of international service role. After a brief review of international modes of service exchange, we will therefore then consider how they may also apply at regional and local levels.

Forms of international service exchange

The growth of international trade and investment has probably been the most important economic development since the Second World War. It is widely credited for the generally high rates of economic expansion during the period. The most successful national economies have been those, such as Germany, Japan, Taiwan or Malaysia that expanded their goods trade with the rest of the world most rapidly. The General Agreement on Tariffs and Trade (GATT) has sought to reduce restrictions on trade. Significantly, service trade became a major issue for the first time between 1986 and 1993, under the Uruguay Round of negotiations. The development since the 1970s of major trading blocs, such as the European Union (EU), has also increased exchange between the member countries, including service trade (Ochel and Wegner, 1987; Cecchini, 1988; Begg, 1989; Bressand and Nicolaidis, 1989; Buiges et al., 1993).

The major participants in international service trade are the developed countries (Box 9.1). The clearest beneficiaries are 'world cities' (Friedmann and Wolff, 1982; Friedmann, 1986), such as New York, London and Tokyo, or large 'gateway cities' (Drennan, 1992), which in the US also include Chicago, San Francisco and Los Angeles. These specialize in financial services to serve national and international needs. Some business consultancy services are traded internationally, but they are often also involved in complex patterns of service exchange with clients who trade, in addition to what is measurable in trade statistics (Clairmont and Cavanagh, 1984; Daniels et al., 1988; Perry, 1990b; Rimmer, 1991).

Services, as part of 'invisible' trade between countries, are routinely measured by the International Monetary Fund (IMF), GATT and the United Nations Conference on Trade and Development (UNCTAD) (Box 9.1). These data also include private and intergovernmental capital transfers and payments (Shelp et al., 1984; Riddle, 1986). Once these are excluded, estimated service trade, in the form of the sale by suppliers in one country of services to clients in another, grew faster than trade as a whole during the 1980s, in business services, finance and transport (Box 9.2). The measured volume of international service trade has nevertheless not generally been sufficient to compensate for significant losses of 'visible' goods trade by major industrial economies. Tradable services thus appear to be a small and specialized segment, in which most of the major economies have at most only a small surplus.

Measured trade statistics, however, do not tell the whole story; they under-represent international service exchange. Other forms of transaction are as significant as direct trade. These are illustrated in Fig. 9.1 (Sapir, 1993). Service delivery generally depends on some form of contact between provider and user. The figure classifies international

Box 9.2: Regional trade in services

Interregional service trade is rarely directly measured. Stabler and Howe's pioneering study of evidence from four western Canadian provinces is thus particularly significant (Stabler and Howe, 1988). It was stimulated by contradictory evidence from previous work, coloured by traditional attitudes to the supposed tradability of services. Some had argued that services cannot be traded to initiate regional growth, while others, including Stanback and Noyelle, assumed this could happen only from major cities. Stabler and Howe asked why service employment had grown to such an extent everywhere, including in many non-metropolitan, small town and city areas. Did this mean that the export potential of these areas had shrunk, as the share of service activities increased, or were their services in fact increasingly involved, directly or indirectly, in trade? Evidence from Beyers and Alvine (1985) in the Puget Sound region of Washington State, around Seattle, suggested the latter conclusion.

The measurement of local service exchanges and interregional trade is, of course, difficult, but they can be estimated in broad terms where suitable input–output data are available. These trace money exchanges between different sectors within an economy, thus accounting for all the inputs, including services, required to produce goods and services for final use, at home or in export markets. At the national level, many studies have shown that dependence on services to support exports from other sectors grew rapidly during the 1970s and early 1980s (Momigliano and Siniscalco, 1983). In the case of Canada (which has particularly good data), it has also been demonstrated that, during the 1960s and 1970s, changes in the productivity of service inputs, such as transportation, communications and financial and business services, had more impact on the final productivity of some client sectors, such as construction, paper, metal manufacturing and food and drink production, than productivity improvements within these sectors themselves (Postner and Wesa, 1984; Wood, 1988).

Stabler and Howe analysed regional input–output data for the Canadian provinces for the 1970s, and showed quite dramatically how services contributed to export growth. First, other exports, dominated by primary products, especially to the rest of Canada, increasingly depended on inputs from local services. More strikingly, however, these relatively peripheral provinces also developed directly tradable expertise of their own. This service trade grew even more rapidly with other countries (mainly the US) than with the rest of Canada. Thus, by the late 1970s, service exports constituted 38 per cent of direct exports, and over half the value of all exports, if their embodied value in other products is included. These data do not reveal the whole story of service sector involvement in economic growth and exchange, and we do not know what the results would be for other regions. They nevertheless offer clear evidence for the scale of the local service contribution to regional exports, from a region with a supposedly resource-based economy. Their results have since been supported by other types of study in other regions (Keil and Mack, 1986; Illeris, 1989b; Gilmer et al., 1989; Gilmer, 1990; Harrington et al., 1991; Browne, 1991; Michalak and Fairbairn, 1993).

transactions in commercial services according to the terms under which such contacts take place. Four dominant cases are identified.

'Type 1' transactions: no international movement of provider or user

This is classical direct trade, dominated at the world scale by material goods exchange. In view of the nature of service transactions, and their reliance upon direct contact between provider and user, it is hardly

Movement of provider

		Low	High	
Movement of user	Low	**Type 1** Wholesale banking Wholesale insurance Telecommunications	Temporary **Type 3** Road transport Business services Airlines Construction	Permanent **Type 4** Retail banking Retail insurance Business services Distribution
	High	**Type 2** Hotels Distribution		

Figure 9.1. Typology of international service transactions

surprising that direct service trade is small. Modern tele-communications, however, have supported significant growth, most obviously in financial services, transmitting both monetary resources and market intelligence around the world (Moss, 1987a,b,c; Warf, 1989, 1991; Moss and Brion, 1991). Its development, however, may still be limited since other forms of exchange, especially under foreign direct investment ('Type 4' transactions), are often more attractive.

Service skills can also be traded in another form; embodied in other traded commodities, especially goods. The growing 'expertise-intensiveness' of high-technology and expertly designed and marketed goods means that service inputs account for a rising share of earnings from goods trade. Although this often remains hidden or ignored, the relative quality of such service inputs between nations is increasingly critical to the competitiveness and success of goods trade (Grubel, 1988).

'Type 2' transactions: the user moves to the source of the service

Modern transport innovation has encouraged the international mobility of people. The best example of this trade is recreational tourism. Mobility to some countries has also increased for health care and education, and to international congresses and conventions.

'Type 3' transactions: the provider moves temporarily to the user

Service expertise is increasingly conveyed through the temporary movement of personnel around the world to serve overseas clients. International labour migration, mainly of low-skilled workers from poor countries, has become a major feature of the modern world economy, with the repatriation of earnings now a significant source of income for countries, such as the Philippines, Egypt or Turkey. High-level skills are also exported from developed countries, however, including key management, engineering, computer, sales, and other consulting staff, often between the branches of multinational companies or other international organizations (Frobel, 1980; Salt, 1984, 1992; Atkinson, 1989; Findlay and Gould, 1989; Johnson and Salt, 1990; Beaverstock, 1990). These create overseas earnings for the 'home' organization and sometimes also for the individuals involved.

'Type 4' transactions: the provider moves permanently to the user

Foreign direct investment (FDI) 'is the most common form of international service competition, enabling frequent and close interaction between buyers and sellers' (Sapir, 1993). In many cases, the decision to invest in overseas supply capacity by multinational business, financial or distribution corporations is undertaken deliberately as a substitute for direct trade. This enables them to attune their services to national needs, regulations and customs which still vary widely for services. They may either establish joint ventures with local partners, or set up their own branches or subsidiaries. Earnings may be reimbursed to the 'exporting' country directly for services rendered through overseas branches or subsidiaries, or through subsequent profits returned on the investment to shareholders and banks.

The purpose of this strategy is usually to establish close contact with clients and to keep control over the ownership and quality of service delivery. Many services depend for their success on the reputation of providers and the trust placed in these. A presence in foreign markets is therefore necessary to establish them. Limitations are placed by many countries on the establishment of foreign service companies, especially in financial and professional markets, but these are being swept aside within Europe, as a result of EU policy, and reduced world-wide, following the 1993 GATT agreement (Nicolaides, 1989; Wood, 1991b; O'Farrell et al., 1994).

Intergovernmental exchanges

Finally, to these generally private sector exchanges should be added intergovernmental payments. Although governments may pay for the exchange of expertise in any of the ways summarized above, they also engage significantly in a 'reverse' form of Type 3 transactions, in which the provider pays for the use of services elsewhere in the world. The major examples in recent years have been in defence-related spending, especially by the US, and services associated with particular forms of financial exchanges, such as loans and development aid (Office of Technology Assessment, 1987; Feketekuty, 1988).

Regional exchange

If we transfer our focus from the international to the regional and local scales, we are now concerned with economies that are usually smaller than nation states, and always more open in their external transactions. The balance between the various wealth-creating roles of services at these smaller geographical scales would therefore be expected to be different in various significant ways.

'Type 1' transactions: regional and local sources of business services

At smaller scales, the share of service inputs purchased 'locally' would be expected to be progressively lower, since the capacity to provide for a full range of specialist needs declines, for example in financial, business and administrative services. It is also easier to import these over relatively short distances from outside, using communications technology or the movement of key workers ('Type 3' below). The 'tradability' of services within countries, with their relative uniformity of customs and regulations, is therefore greater than between them (Box 9.2). This has favoured the concentration of specialist sources into core regions (Michalak and Fairbairn, 1993; Moulaert and Gallouj, 1993). The impact of these trends on wealth-creation in each region or locality depends on its changing balance of service exports and imports (Harrington et al., 1991). Generally, areas with a net export balance are probably becoming fewer, as expertise and infrastructure concentrate into centres offering agglomeration economies.

This is not an inevitable trend, however. In more geographically and administratively dispersed countries, such as the US, Germany or Canada, regional service specializations have emerged, based on the modern ease of service communication between complementary centres of expertise. In historically centralized countries, such as the UK or France, however, the dominant capital regions still tend to attract the lion's share of specialized services. The degree of regional exchange thus varies, and is also not static. Even in the UK and France, some quite high-level services are moving out of the capitals to provincial cities (Marshall et al., 1988; Leyshon et al., 1988; Leyshon and Thrift, 1989; Aksoy and Marshall, 1992). New 'core' regions may also emerge, especially from industrial and trading success. Regions such as the electronics-based 'Silicon Valley' in California, the flexibly organized regions of the 'Third Italy', or the engineering-based Baden Württemberg in Germany may develop significant net exports of specialist services, initially embodied in other local activities (Benko and Dunford, 1991). The same may happen, of course, with regions specializing in financial, business or transportation services.

The local wealth-creating contribution of services 'embodied' in other, especially goods trade is also likely to be lower at regional and local scales (Box 9.2). It has often been suggested that a lack of local support services adversely affects economic development. Certainly Hansen (1990) concludes that producer services have the capacity to raise regional productivity and growth rates in per capita income. Service-rich regions, with their high incomes, also seem to attract other types of manufacturing, property, infrastructure and consumer investment. Elsewhere, branches of large companies, in particular, often rely on expertise communicated from their headquarters, thus discouraging the development of local business services. This may undermine local service availability to other firms (Marshall, 1983, 1985; Wood, 1987; MacPherson, 1988, 1991; O'Farrell and Hitchins, 1990a, 1990b, 1992; O'Farrell *et al.*, 1993). The implications of this depend on the relative quality of such services; there can be little advantage in using second-rate local advice, when the best can easily be acquired from elsewhere. Local availability seems to be most critical, however, for small- and medium-sized businesses. These often lack the expertise and contacts to exploit more distant sources of service support for technical, management or marketing innovation. In some countries, regional policies have been initiated to help them gain such assistance, including the 'Enterprise Initiative' in the UK, and the establishment of 'Business Service Centres' in Italy. Other forms of local infrastructure, planning and administration, including business information and advisory services, also attempt to influence location decisions (Bennett and McCoshan, 1993; Bennett and Krebs, 1994), but they are nowadays so widespread in the advanced economies that their differential impacts are generally marginal.

'Type 2' transactions: attracting people

Mass interregional tourism of course predated similar international movements by at least a hundred years. It was a product of the railway age, in the same way that international tourism has more recently arisen from air transport. Today, few regional or local economic development programmes lack a tourist policy component, mainly to attract automobile-based travellers to experience scenery, heritage or opportunities for entertainment.

Localities may offer other services to people from outside, and many have developed distinct strategies to do so. Universities and colleges have long dominated some world-famous towns and cities, such as Bologna (Italy), Cambridge (England; Massachusetts, US), Oxford (England), St Andrews (Scotland), Princeton (New Jersey, US) and Heidelberg (Germany), and the modern expansion of higher education has broadened their role in regional and local economic development. Similarly, major clinics and hospitals, sometimes associated with historic spas, are now powerful magnets both for medical treatment, and various types of ancillary service. The Mayo Clinic in Rochester, Minnesota, is the modern model of such development.

Modern consumer mobility, however, has promoted new types of opportunity to attract people and their spending power to localities and regions, even from abroad. Among the dominant symbols of this trend are the convention and exhibition centres and arts complexes which were widely developed during the 1980s. These often attract business, as well as tourist and consumer spending, also sometimes associated with the large-scale shopping, entertainment and office developments described in Chapter 7. Another form of consumer and touristic experience has developed around the modern sports stadium, often also presenting other forms of entertainment. Most visibly, a calendar of major sporting events has developed, associated both with particular places, and regular movement between localities which vie to stage them. Augusta, Georgia (golf), Wimbledon (tennis), Wembley (soccer) and Indianapolis and Le Mans (motor racing) are among the fixed points in this calendar. Mobile events include the four-yearly Olympic Games and soccer World Cup, and the annual Super Bowl (American football) and World Series playoffs (baseball). Staging these enables communities not only to attract consumers, but also to project their image nationally and world-wide through TV and press coverage to potential investors (Bale, 1989).

'Type 3' transactions: business travel

Like tourist and consumer movement to such events, the level of mobility of specialist service labour is also much higher between regions than internationally. If service expertise cannot be directly commun-icated to other regions, as in a 'Type 1' transaction, experts may easily travel to customers from their home region (Johnson and Salt, 1990). Such regions therefore benefit through salary or fee payments. This practice, of course, is now a characteristic aspect of modern business, and also of public administration, encouraged by the ease of passenger transport. Even more than remotely communicated expertise, it tends to focus on central regions, with junior management from regional outposts travelling to meet centrally based headquarters staff (Marshall, 1978). The latter are usually based at locations with maximum accessibility to national and international markets.

'Type 4' transactions: interregional and local service investment

At the regional as at the international scale, investment by outside companies embodies many service skills which earn wage and investment income for the headquarters region. The impacts on trade and earnings patterns at this scale are, however, somewhat different. Internationally, we have seen that overseas investment by service companies is, to some extent, a substitute for the direct communication of service skills from the 'home' country. It is usually undertaken, at significant cost, to ensure a quality of service delivery to clients which cannot be achieved from a distance. Local commercial, regulatory and cultural circumstances also have to be accommodated in delivering any service abroad. The higher the quality of a service, the more likely it is that offices will be established there to replicate the service offered in the home country, either through joint ventures with local firms, or by establishing branches or subsidiaries (O'Farrell et al., 1995). Conse-quently, even though some profits are also likely to be repatriated, wealth creation in receiving regions may be considerable.

In contrast, investment in other regions within the same country, including taking over established businesses, is more likely to increase interregional trade and therefore concentrate wealth into the core or 'exporting' region. It tends to enhance local specialization, intensifying the spatial division of labour and the resulting exchange of both goods and specialist services between regions. Higher-level service skills generally remain concentrated into headquarters regions, with routine technical and administrative functions perhaps more widely dispersed (Daniels, 1986a; Coffey and Polese, 1987, 1992; Green and Howells, 1988; Morris, 1988; Goe, 1990; Martinelli, 1991). This is partly because of the greater ease of communication and travel within countries, compared with between them. Compared with international markets, cultural, language, legal and commercial conditions between regions within a country also show much greater homogeneity.

Thus the attraction of service investment to regions or localities is more likely to involve specialized or subsidiary aspects of activity, rather than the higher-level functions found in international investment. Much such investment has, of course, been associated with the movement of 'back-office' functions from more expensive core areas (Glasmeier and Borchard, 1989). The precise activities attracted to particular sites differ between cases, however, and there is evidence that progressively higher-level activities are now being decentralized from major cities, under the influence of more devolved forms of organization and increasingly sophisticated communications technology. Employment levels in core region head offices, even while they retain strategic functions, are also being widely reduced (Aksoy and Marshall, 1992). Thus regional income from some service investments may be growing, even while others are being focused elsewhere (Bailly et al., 1987; Kirn, 1987; O hUallachain, 1989, 1991; Marshall and Jaeger, 1990; Beyers, 1991).

Intragovernment exchanges

Non-market, government-initiated service exchanges are also more important at the regional scale. The most obvious of these arise from national government administration and services operating from particular regions, usually in the capital. In some cases, as we have seen, national functions have been explicitly dispersed to support regional development policies (Chapter 8). In others, the regional effect is more incidental, though no less powerful, as has been seen for the distribution of defence functions (Malecki, 1984; Lovering and Boddy, 1988; Breheny, 1988). Even more pervasively, very large transfers of 'invisible earnings' arise at the regional scale through the income redistribution effects of taxation and social security payments, as well as from patterns of government investment including regional or urban subsidies. These favour the poorer regions of many countries, although they have been reduced in scale as a result of the public spending squeezes of the 1980s. Within the European Union, however, national transfers were significantly augmented during the 1980s by the growth of social, structural and regional shifts of resources between the member states, analogous to the interstate transfers common in the US.

225

Regional service specialization

Financial services

As we have indicated, many forms of international and national service transaction have supported specialized regional growth in recent years. Two particular forms of commercial service trade have attracted special attention; financial services, mostly Type 1, and tourism, Type 3. In many countries, they have focused their international wealth-creating capacity into distinctive regions. The financial services manage capital resources, increasingly on an international as well as national scale. They support, and profit from, the circulation of capital, partly as a service to production, but also to make speculative gains for funds that are surplus to direct or immediate productive needs. They thus form a group of financial institutions which provide the means for those who wish to invest money to do so. Such investors may be multinational companies, banks, pension funds, insurance companies, governments or individuals. Investment may be in bank accounts, in currency holding funds, stock exchange equities and related funds, government bonds, futures markets, insurance risks, or in loans for particular projects, arranged as securities through banks, or increasingly through specialist companies (Thrift, 1987, 1990a,b; Marshall *et al.*, 1988; Rajan and Fryatt, 1988; Gentle and Marshall, 1992).

The rapid growth in demand for financial service expertise has followed the expansion of international capital exchanges since the early 1970s. Warf (1989, 1991) indicates that this was especially stimulated by the spread of instantaneous communications and powerful computer data processing during the 1980s. The changes were driven, however, by the competitive deregulation of international financial transactions, especially in the US and the UK, mainly during the 1980s (Thrift, 1987; Leyshon and Thrift, 1992). This allowed capital movements which had formerly been constrained by national controls and regulations to move freely between the major industrial centres of North America, Europe and the Far East. Multinational corporations, banks and other institutions with large reserves thus had open channels for moving their funds around the world in search of profitable short-, medium- and long-term investments (Thrift 1987; Thrift and Leyshon, 1988; Daly and Stimson, 1994).

Paradoxically, the ease and volume of exchange has focused business into a few countries, and a few centres and companies within these (Marshall *et al.*, 1992). They possess the essential expertise required to interpret and direct the daily volumes of capital exchange, exploiting intelligence networks, close business contacts and experience of a wide variety of financial functions. The data processing and deal-making services thus centre on three major world cities: New York, Tokyo and London. Other centres, such as Paris, Frankfurt, Amsterdam, Milan, Hong Kong, Singapore, Los Angeles and Chicago, play significant, but secondary roles in the international market (Wheeler and Dillon, 1985; Wheeler, 1986; Harvey, 1989a; Thrift, 1990 a,b). The dominance of New

York and Tokyo reflects the status of their domestic markets. London maintained its role as the main European financial centre, in spite of the relatively weak UK economy, by adapting its traditional world trading role (Budd and Whimster, 1992). Its financial sector was less regulated than rival European centres during the 1980s, especially after the 'Big Bang' of 1986. Traditional structural divisions and restrictive practices were then removed to allow large companies to diversify across many markets. The breadth of expertise offered by London-based, although often non-UK-owned, financial institutions was based on the world network of trading and banking contacts; the UK's tradition, like that of the US, of independent corporate accounting; the scale of its managed insurance and pension funds; and the regular experience it had gained of corporate restructuring, including acquisition and merger processes.

In all three world centres, the growth of international finance spilled over into domestic financial and property markets, in the search for new forms of profitable investment (Pryke, 1991, 1994; Gentle and Marshall, 1993). A consumption boom was fuelled both directly through the increasing wages of financial staff, on ever-increasing 'golden hellos' and 'golden handshakes'; and indirectly through the deregulated expansion of consumer credit (Thrift and Leyshon, 1992; Gentle, 1993). The negative, polarizing impact of this growth on the employment structure of the cities, is well documented by Sassen (1989, 1991; see Chapter 5). The integration of the financial sector with other speculative sectors added to the volatility of national economies as a whole, and ended in the crash in the late 1980s, as interest rates increased to control rising inflation.

There is little doubt that the events of the 1980s created spectacular growth in and around three major world cities, (for example, see Frost and Spence, 1993) and made a substantial contribution to the growth or revival of many second- and third-order centres. Financial services show, in their most advanced form, the potential of regulatory and technical change in services to integrate international exchanges, to create new markets, and to focus the consequent growth into particular regions. A question remains over whether the dramatic events of 1980s reflect a one-off explosion, released by the deregulation of financial markets in many countries (Molyneux, 1989; Johnson, 1991), or a longer-term change associated with a round of innovation in information and communications technology-based services (Begg 1992; Barras, 1985; Miles, 1993). In the late 1980s and early 1990s, sharp reductions in financial service employment were being experienced in more routine employment in capital markets (Cox and Warf, 1991) and in banking and insurance functions, often aligned to national markets (Leyshon and Thrift, 1993), but the vitality of international capital exchanges seemed to be reviving, as recession eased.

Tourism

Holiday-making and tourism are among the most distinctive manifestations of modern consumerism, reflecting not only growing

227

spending power but also greater opportunities to express personal preferences and tastes (see Chapter 7). A powerful 'industry', or rather group of industries, including transportation systems, hotel and restaurant chains, and retailing and media networks, has developed to supply tourist opportunities. These are focused most intensely into regional complexes where sun, sand and scenery are combined, but many large cities also have developed tourist sites (Mathieson and Wall, 1982; Murphy, 1985; Perry, 1986; Pearce, 1987; Mills, 1989; Batty, 1990; Johnson and Thomas, 1990, 1992; Williams and Shaw, 1991; Williams *et al.*, 1989; Ward, 1991; Law 1992). Holidays nevertheless have a widely varied economic and social significance, which gives many locations at least some potential for tourism. Tourist preferences differ widely according to their degree of independence (e.g. on package tours, as mountain campers), purpose (seeking sun, adventure, cultural experiences), favoured types of place (Florida, the Himalayas), forms of accommodation (five star hotel, self-catering), or typical group behaviour (family, 18–30, climbing club). Although variably available to different social groups in high-income societies, holidays are generally accepted to be an essential component, if not a right, of modern life.

Tourism has always essentially required a population with sufficient income, leisure time away from work, and access to particular modes of travel. Before the railway age, these were confined to élite groups, for example taking the European grand tour, or the waters and social life of spa towns. Such élites later pioneered more exotic world travel, through ocean cruises. Mass holiday-making depended on mass transportation and regulated working conditions, enabling the middle and later the working classes to travel to seaside or mountain resorts, often associated with rail links to major population centres. These were made to be as 'extraordinary' as possible, compared with everyday working experience of industrial towns and cities (Urry, 1988). Such primarily national patterns of holiday-making reached their apogee in Europe in the 1950s.

Since then, as earlier in the US, these patterns have been transformed by increased incomes, automobile travel, the growing range of holiday choices, and especially by mass air transport. Tourism, perhaps more than any other form of consumption, now depends on the creation and exploitation of consumer myths and fantasy. According to Urry, modern forms of 'post-(mass) tourism', offer a multitude of choices, making almost any part of the world familiar through media images of other places. Visiting such places thus becomes more an exercise in verifying the truth of images, than seeking a new reality, whether these are perfect weather or the thrill of the mildly exotic (Urry, 1988, 1990a,b; Britton, 1991).

Tourism, like other consumer products, now offers a wide range of competing choices, many of which have become highly commercialized, so that tourist regions are engaged in intense competition with each other, on an international scale (Thompson, 1990). Broadly the same pattern of growing capital domination is found in the operation of the commercial tourist sector, as was described for retailing in Chapter 7.

Major leisure-based groups, with interests in hotels, catering, travel and entertainment now serve mass markets. These are less dominant in domestic markets, where they have had to adjust to growing competition from international and 'self-service' holidays, and have specialized more into various forms of provision, including short breaks, tours and 'themed' holidays.

At the international scale, however, commercial tourism extends across continents, providing major sources of income even for developed countries, such as Italy, Spain and Greece, as well as for small developing countries in the West Indies, Africa and Asia. These developments are associated with concentration of capital, the emergence of diversified leisure-based companies, sometimes within wider corporate conglomerates, and often commercially associated with particular airlines. The most significant current development in the supply of holiday opportunities is probably market differentiation, and the emergence of specialist sectors. These focus, for example, on winter holidays, particular age groups (18–30; over 50); self-catering holidays (camping, villas); culturally oriented holidays, coach tours or cruises; adventure holidays, long-haul travel or visits to particular countries.

The implications of tourism for regional wealth creation are similar to other forms of consumption discussed in Chapter 7. Tourism supports generally poorly paid and seasonal labour, and much of the commercial benefits accrue to the international companies controlling the various aspects of service provision. Sometimes, public investment, especially in poor countries simply subsidizes their activities (Lea, 1988; Mansfield, 1990). In addition, the tourist industry is highly competitive, cyclical, and at times extremely volatile, 'depending on many external factors over which a destination area has little or no control. Success can be influenced by weather, changing consumer tastes, demographics, economic cycles and government policy' (Murphy, 1985; quoted in Griffiths, 1994), not to mention international terrorism, and other forms of conflict.

The contribution of tourism to trade and local development can, however, be significant. For some, especially remote areas, it may be one of few options available for significant economic development, including investment in facilities and infrastructure. It also builds and markets images of areas which may provide a wider stimulus to local development, including residential and other commercial activities. The money which tourists spend at theatres, museums and concerts may also help to make these activities viable. Perhaps of all service activities, however, because it is, literally, 'invasive', tourism needs to be developed in association with other economic functions, and related to wider programmes of social development (Sinclair and Page, 1993). It may thus contribute valuable opportunities for modern consumption and infrastructure investment, especially in developed regions, where it can also be associated with wider economic development strategies.

Services and local economic development

The service contribution to wealth creation is obviously significant at all geographical scales of analysis, but depends on a complex network of relationships and exchanges with other economic functions, at local, regional and national levels. Far from being predominantly locally based, most services in local economies, and many in regional economies, come increasingly from outside sources. This results from higher private and public sector dependence on service trade, lower use of local services by other local agencies, the greater mobility of consumers, especially tourists, and also of key service workers, and growing external control of service businesses and agencies.

In local labour markets in the UK, the most geographically uneven service activities in 1989 were public administration and defence, and hotels and catering (Townsend, 1991). The processes of geographical concentration are thus at least as powerful in the public sector and tourism as in the financial and business services. The net balance between service imports and exports in particular places thus reflects the changing structure of public administration, the incidence of tourist attractions, and the geography of corporate control, focusing or dispersing business functions in and around core regions. Government-based transfers of wealth have balanced this concentration in the past, but their compensating impact is generally being reduced.

On the other hand, communications improvements and corporate and state devolution probably now offer some regions the opportunity to attract more service investment, even if not at the highest strategic level. Local development strategies need to grasp these opportunities. Success is most likely by tapping developments in various sectors. The financial and business services have attracted attention in recent years, but their employment growth will be more selective in future, and geographical dispersion will take place mainly to other cities, and localities accessible to core regions. Other places may develop or sustain strengths in government-supported education, health or administrative functions. Tourism is also one option for a widening range of places. Elsewhere, there may be potential for convention centres, regional focuses of retail consumption and recreation, or venues for major sporting events. And, of course, manufacturing, including high technology and consumer-oriented production, must be a significant component of any development programme where it can be fostered or attracted. Local interdependence between these activities may not be great, although business services may be attracted to regional assemblages of manufacturing, public sector, tourist and consumer clients. Diversity is important, however, at least on a regional scale, because it supports the vital community, communications and transport infrastructure needed to sustain the modern economic attractiveness of any place.

A 'service-informed view' of modern economic restructuring thus does not suggest that any single sector provides a panacea for regional or community wealth creation. Manufacturing alone is never enough, and in most places is unlikely to support much employment. In spite of their recent growth, financial and business services can make no more than a selective contribution outside core regions. Particular places may attract consumer spending from a wider region, but this may be at the expense of other nearby locations. Tourism and major events or spectacles are usually seasonally variable, offering only insecure and low-income jobs. The economic impacts of one-off events are also often small, at least in isolation. Together, however, some combination of these, serving different segments of wider regional and national markets, may sustain the critical levels of infrastructure, quality of community life and attractiveness to investment required to create successful economic development in the late industrial era. The modern service-based economy may not offer the obvious, focused source of regional and local wealth that smokestack factories did in the past. Instead, many functions need to be combined, linking localities to sources of outside investment through the many economic relationships which sustain the various types of service.

The integration of social and economic trends in local economies

We have presented the economic logic behind developments in service activity and employment at the local and regional scale, linked to international and national developments. At the local scale, however, we saw in Chapters 5 and 7 that economic changes must be assessed in relation to wider processes of social change and their effects on domestic life and patterns of consumption. At the community level, recent decades have seen more diverse household structures and relationships, new gender roles, greater inequality and diversity in spending patterns, the development of new types of informal work and 'self-service', the rise of new consumer services, especially related to new manufactured commodities, changing attitudes to consumer credit, and very different expectations from public services. These diverse trends have major implications for how local services, and the employment they provide, affect local communities, and the extent to which they meet individual needs.

The growth of service employment, in the recent past, and for the foreseeable future, seems likely to combine with these changes (to different degrees in each country), to create pressures for greater social and spatial polarization in the advanced industrial economies.

Social polarization

The contrast between a growing demand for highly skilled, professional work, especially associated with new technologies, and the expansion of 'casual' low-skilled service jobs looks set to continue. While there will

231

be less long-term employment for males, as a result of decline in blue collar production jobs, women will have to go out to work more often, usually in part-time service employment. Traditional gender relations may thus continue to change. Women will, nevertheless, continue to supply the bulk of domestic labour. Public sector rationalization predicated on shifting caring functions into the home and the voluntary sector will add to the tensions arising from this. Women will also be under pressure from technical changes, rationalizing routine clerical and administrative jobs. At the same time, the trend towards more women taking professional jobs, supported by female domestic staff will continue, increasing the diversity of female labour.

Older people will find it difficult to get jobs when so many employers in consumer services want younger employees who will convey the 'right image' to their customers. In European economies, at least, jobs lost in manufacturing look unlikely to be fully replaced by increasingly capital-intensive services. As well as more unemployed, there may be more homeless, travelling people, and groups or individuals seeking casual work. Thus, communities will become more economically and socially divided, and many will become dependent on established and new informal modes of provision to 'get by'.

Spatial polarization

Whatever success local economic development strategies may have, if high-level jobs in business, financial, consumer and public services remain relatively concentrated into fewer regions and localities, many communities, including old industrial and mining regions, declining agricultural and rural areas, and inner cities, will be left mainly with more routine, part-time and lower-paid commercial and public sector jobs, as well as higher levels of unemployment. Cumulative processes will also create growing inequality of service provision, with a relative decline of public service support, and the switch to private provision by the better-off.

Even within wealthy communities, we have seen that the growth of high-level service functions often depends on the support of poorly paid, insecure employment, in many ancillary jobs. This relationship has been obscured by the influence of the growing and re-forming 'service class', displacing local industrial capitalists or landed interests. As Crang and Martin demonstrate in the reading in Chapter 5, the rise of service class values, although by no means uniform, are tied to particular, largely privatized forms of consumer aspiration, lifestyle and career. This tends to create wider social contradictions and conflicts, especially at the local level. These may find expression in resistance to planned economic development on environmental grounds from the better off, when the needs of other groups for jobs remain unsatisfied. It may also lead to pressure for improvements in local leisure and recreation provision even as social and welfare services are stretched.

Important areas of social, rather than specifically economic, change are thus implicated in service developments. Service change is much more comprehensively engaged with modern social change than manufacturing. Services are locally produced in diverse, interconnected ways, and also combine various formal, informal and commodified modes of delivery. As important as the priorities of private capital in service development, and the supportive role of the public sector, is the continuing capacity of individuals and other organizations to opt for alternatives, in acquiring the basic needs of life including 'self-service' and informal sources of support within communities. But whatever strategies are employed by individuals and families, and whatever complex shifts occur between the formal and informal economy, economic realities in the formal economy are likely to continue to dominate future patterns of local service provision and opportunity.

Further discussion and study

The best way to discuss and develop the ideas presented in this final chapter, and in the book more generally, is through study of particular types of economy, employing desk research or local field investigation. This should review the structure of the economy at large, including manufacturing and other materials processing functions, and the character of private and public sector service change within this. Activities should be identified which involve exchanges with other places, either directly through trade, labour mobility or investment, or indirectly, as inputs to other exported goods or services. Each form of 'export', of course, is balanced by similar imports for local consumption from elsewhere.

The direct measurement of these exchanges, and the investment and employment they support, is seldom possible, but a general assessment of the balance of exchange is worth trying. For example, if employment in a particular activity is significantly greater than the regional or national average (i.e. the 'location quotient' is > 1), the area is probably a net exporter of that service. Examples of two types of area are presented here, to illustrate the questions which might be addressed.

World cities; critical service-based hubs in the modern economy

Examples are New York, Tokyo, Hong Kong and London.

1.   How far did these cities first develop as bases for manufacturing production and trade, providing a local market, a pool of manual labour and good transport facilities?
     How significant is manufacturing in the economy today? What sorts of job have been lost in manufacturing, in terms of gender, income, skills and conditions of employment? Why has manufacturing declined in many large cities?
2.   How significant are services today, compared with other areas, and the past? What proportion are in 'export' sectors? What sort of jobs are being provided in the service sector? How many are high/low

233

income, full-time/part-time/seasonal, permanent/temporary, for women/men?

3. Look at particular 'exporting' sectors, including finance, business services or national and international government. What links are there between these and other local activities? For example, have they affected local property values, or stimulated local consumer service employment?

4. Consider the role of the city as a centre of consumption. What hinterland region does it serve for retail sales, including those based on daily commuting? What have been the most successful and unsuccessful shopping-based developments? How important is tourism as a source of income? How far does it support other retailing, cultural and entertainment activities?

5. How many are employed in the public sector? What are its distinctive tasks in cities such as this? Where does the money to support them come from? Is there resistance from any local groups to the payment of taxes to support the public sector? Has there been any recent reorganization of the public sector, and what are its implications for the local economy?

6. Is there a corporate complex of head office and business service suppliers at the heart of the metropolitan economy? What is the balance of large and small firms? What business services are exported to other places?

7. To what extent are congestion pressures and high costs a constraint on the economy, and how are industries and services adapting to them, e.g. by *in situ* reorganization or decentralization?

8. What role does the telecommunications infrastructure play in the city? How important are innovations such as teleports or high-quality data transfer facilities to the national and international operations of large business and finance capital?

Guided reading

Daniels P W 1993 *Service industries in a world economy*. Blackwell, Oxford
Daniels P W 1991 A world of services? *Geoforum* **22**: 359–76
Hall P 1984 The world cities, 3rd ed. Weidenfield and Nicolson, London
Hall P 1989 The rise and fall of great cities: economic forces and population responses. In Lawton R (ed) *The rise and fall of great cities*. Belhaven, London: 20–31
Sassen S 1991 *The global city: New York, London, Tokyo*. Princeton University Press, Princeton NJ
Thrift N J, Leyshon A 1992 In the wake of money. In Budd L, Whimster S (eds) *Global finance and urban living*. Routledge, London: 282–311
Warf B 1991 The internationalisation of New York services. In Daniels P W (ed) *Services and metropolitan development. International perspectives*. Routledge, Chapman and Hall, Andover, Hants: 245–64

Less-favoured industrial regions: can the growth of service activities help rejuvenate them?

Examples are North East England, the Ruhr in Germany, southern Italy, Buffalo or Pittsburg regions in the US.

1. What services are represented, and how much growth has occurred? What sort of jobs have been created, and how do they compare with those lost in manufacturing? Who is losing blue collar jobs, and can service growth replace them?
2. In general, what sectors have been relatively more successful in sustaining employment in recent years: new forms of manufacturing; consumer services; publicly funded projects; transport and materials handling; construction; business and financial services?
3. Can there be consumer-led service initiatives when many local incomes are depressed by manufacturing decline? What sort of local jobs, and how many, are still secure and well paid? Can the reorganization and consolidation of retailing into new centres help the area's renewal?
4. Is there any evidence of business services, including materials handling and transport services as well as white collar jobs, developing an export role independently of local manufacturing? Or are they being dragged down by its demise?
5. How far is the region a manufacturing branch plant economy? What impacts do branches have on the local service sector? Is there any evidence of local branch plants becoming more autonomous, developing significant management skills locally?
6. What advisory services are available for firms wishing to invest in the area, from private or public sector sources, and what support is there for land development, training, infrastructure investment, etc.?
7. What role has finance capital played in the region? Is investment mainly from private companies, banks, venture capital agencies or public subsidy? How far are national or international financial organizations active in the region?
8. How dependent is the area on employment in the public sector? Is this being affected by restructuring policies?
9. How far are these old industrial areas peripheral to the economy at large? How important is physical remoteness, and the lack of good transport or telecommunications? What policies have tried to improve these?
10. Identify successful and unsuccessful development projects in the areas. Compare their characteristics and the reasons for the outcomes.

Guided reading

MacPherson A 1991 Inter-firm information linkages in an economically disadvantaged region: an empirical perspective from metropolitan Buffalo. *Environment and Planning* A, **23**: 591–606

Martinelli F 1991 Branch plants and services underdevelopment in peripheral regions: the case of southern Italy, in Daniels P W, and Moulaert F (eds) *The changing geography of advanced producer services*. Belhaven, London: 151–77

Perry M 1991 The capacity of producer services to generate regional growth: Some evidence from a peripheral metropolitan economy. *Environment and Planning A*, **23**, 1331–47

Townsend A 1991 Services and local economic development. *Area* **23**: 309–17.

Bibliography

Abercrombie N, Urry J 1983 *Capital labour and the middle classes*. Allen and Unwin, Winchester, MA

Abler R 1970 What makes cities important? *Bell Telephone Magazine* **49**: 10–15

Adams R, Hamil S 1991 The powers that buy. *Marxism Today*, July: 269–89

Aglietta M C 1979 *A theory of capitalist regulation: the US experience*. New Left Books, London

Aksoy A, Marshall J N 1992 The changing corporate head office and its spatial implications. *Regional Studies* **26**: 149–62

Aksoy A, Robins K 1992 Hollywood for the 21st century: global competition for critical mass in image markets. *Cambridge Journal of Economics* **6**: 1–22

Alexander I 1979 *Office location and public policy*. Longman, London

Allen J 1988a Service industries: uneven development and uneven knowledge. *Area* **20**: 15–22

Allen J 1988b The geographies of service. In Massey D and Allen J (eds) *Uneven redevelopment: cities and regions in transition*. Hodder & Stoughton, London: 123–42

Allen J 1988c Towards a post-industrial economy. In Allen J and Massey D (eds) *The economy in question*. Sage, London: 91–135

Allen J 1992 Services and the UK space economy: regionalization and economic dislocation. *Transactions of the Institute of British Geographers* **17**: 292–305

Allen J, Henry N 1994 Growth at the margins: contract labour in a core region. In Hadjimichalis C, Sadler D (eds) *In, on and around the margins of a new Europe,* forthcoming.

Atkinson J 1989 *Corporate employment policies for the single European market.* Report No. 170, Institute of Manpower Studies, University of Sussex

Bacon R, Eltis W 1978 *Britain's economic problem: too few producers.* Macmillan, London

Bagguley P 1991 Post-Fordism and enterprise culture: flexibility, autonomy and changes in economic organisation. In Keat R, Abercrombie N (eds) *Enterprise culture.* Routledge, London: 151–70

Bailly A S, Maillat D 1988 *Le secteur tertiaire en question: activités de service, developpement economique et spatial.* Editions Regionales Europénnes, Geneve

Bailly A S, Maillat D 1991 Service activities and regional metropolitan development: a comparative study. In Daniels P (ed) *Services and metropolitan development.* Routledge, Chapman and Hall, Andover, Hants: 129–45

Bailly A S, Maillat D, Coffey W J 1987 Service activities and regional development: some European examples. *Environment and Planning A* **19**: 653–68

Bannon M J (ed) 1973 *Office location and regional development.* An Foras Forbartha, Dublin

Barcet A, Bonamy J, Mayere A 1983 *Economie de services aux entreprises.* Economie et Humanisme, University de Lyon, Lyon

Barras R 1983 *Growth and technical change in the UK service sector.* Technical Change Centre, London

Barras R 1985 Information technology and the services revolution. *Policy Studies* **5**: 14–24

Barras R 1986 New technologies and the new services. *Futures* **18**: 748–72

Barron I, Curnow R 1979 *The future with microelectronics.* Frances Pinter, London

Bateman M 1985 *Office development: a geographical analysis.* Croom Helm, London

Batty B 1990 Tourism and the tourist industry. *Employment Gazette.* **98**: 438–47

Baumol W J 1967 Macro-economics of unbalanced economic growth: the anatomy of urban crisis. *American Economic Review* **LVII**: 415–26

Baumol W J 1985 Productivity policy and the service sector. In Inman R P (ed) *Managing the service economy.* Cambridge University Press, Cambridge: 301–17

Beaverstock J V 1990 New international labour markets: the case of professional and managerial labour migration within large chartered accountancy firms. *Area* **22**, 151–8

Begg I 1989 The regional dimensions of the '1992' proposals. *Regional Studies* **23**: 368–76

Begg I 1992 The spatial impact of completion of the EC internal market for financial services. *Regional Studies* **26**: 333–47

Begg I 1993 The service sector in regional development. *Regional Studies* **27**: 817–35

Bell D 1973 *The coming of post-industrial society.* Heinemann, London

Beniger J 1986 *The control revolution: technological and economic origins of information society.* Harvard University Press, Cambridge, MA

Benko G, Dunford M (eds) 1991 *Industrial change and regional development: the transformation of new industrial spaces.* Belhaven Press, London

Bennett R J, Krebs C 1994 Local economic development partnerships: an analysis of policy networks in EC-LEDA local employment development strategies. *Regional Studies* **28**: 119–40

Bennett R J, McCoshon A 1993 *Enterprise and human resource development.* Paul Chapman, London

Berry B L 1966 *The geography of market centres and retail distribution.* Prentice Hall, Englewood Cliffs, NJ.

Bertrand O, Noyelle T J 1988 *Corporate strategy and human resources: technological change in banks and insurance companies in five OECD countries.* OECD, Paris

Beyers W B 1989 *The producer services and economic development in the US: the last decade.* Final report for the US Department of Commerce. Department of Geography, University of Washington, Seattle.

Beyers W B 1990a Changing business practices and the growth of producer services, paper presented at the Association of American Geographers, Toronto. Available from the author, Department of Geography, University of Washington, Seattle WA

Beyers W B 1990b Contrasts in producer services in the USA and UK. Paper for the Fulbright Colloquium, *Buying back America: UK capital in the US service economy*. Vale of Glamorgan (Available from the author, Department of Geography, University of Washington, Seattle)

Beyers W B 1991 Trends in the producer services in the US: the last decade. In Daniels P W (ed) *Services and metropolitan development: international perspectives*, Routledge, Chapman and Hall, London: 146–72

Beyers W, Alvine M J 1985 Export services in post-industrial society. *Papers of the Regional Science Association* **57**: 33–45

Bhagwati J 1987 International trade in services and its relevance for economic development. In Giarini O (ed) *The emerging service economy*. Pergamon, Oxford: 3–34

Bird J 1971 *Seaport and seaport terminals*. Hutchinson, London

Blackaby F (ed) 1979 *De-industrialisation*. Heinemann, London

Blackburn P, Coombes R, Green K 1985 *Technology, economic growth and the labour process*. Macmillan, London

Blois K J 1985 Productivity and effectiveness in service firms. In Foxall G (ed) *Marketing in the service industries*. Cass, London: 49–60

Blomley N K 1987 Retail regulation in England and Wales: the results of a survey. *Environment and Planning A* **19**: 1399–406

Bocock R 1993 *Consumption*. Routledge, London

Bondi L 1990 Progress in geography and gender: feminism and difference. *Progress in Human Geography* **14**: 438–45

Bondi L, Peake L 1988 Gender and the city: urban politics revisited. In Little J, Peake L, Ritchardson P (eds) *Women in cities*. Macmillan, London: 21–40

Boyer R 1986 *La theorie de la regulation: une analyse critique*. Editions la Decouverte, Paris

Bradford M, Burdett F 1989 Spatial polarisation of private education in England. *Area* **21**: 47–57

Braverman H 1974 *Labour and monopoly capitalism.* Monthly Review Press, New York.

Breheny M J (ed) 1988 *Defence expenditure and regional development.* Mansell

Bressand A, Nicolaidis K (eds) 1989 *Strategic trends in services: an enquiry into the global service economy.* Harper & Row, New York

Brewer A 1984 *A guide to Marx's Capital.* Cambridge University Press, Cambridge

Britton J N H 1974 Environmental adaption of industrial plants: service linkages, locational environment and organisation. In Hamilton F E I (ed) *Spatial perspectives on industrial organisation and decision-making.* Wiley, New York: 363–90

Britton S 1990 The role of services in production. *Progress in Human Geography.* **14**: 529–46

Britton S 1991 Tourism, capital and place: towards a critical geography of tourism. *Environment and Planning D: Society and Space* **9**: 451–78

Bromley R D F, Thomas C J 1993a The retail revolution: the carless shopper and disadvantage. *Transactions of the Institute of British Geographers* **18**: 222–36

Bromley R D F, Thomas C J (eds) 1993b *Retail change: contemporary issues.* UCL Press, London

Brown S 1987 Institutional change in retailing: a geographical interpretation. *Progress in Human Geography* **11**: 181–206

Browne L 1991 The role of services in New England's rise and fall: engine of growth or along for the ride. *New England Economic Review* **July/August**: 27–44

Browning H L, Singelmann J 1975 *The emergence of a service society.* National Technical Information Services, Springfield, Virginia

Brusco S 1982 The Emilian model: productive decentralisation and social integration. *Cambridge Journal of Economics* **6**: 167–84

Bryson J, Keeble D, Wood P A 1993 Business networks, small firm flexibility and regional development in UK business services. *Entrepreneurship and Regional Development* **5**: 265–77

Buck N 1985 Service industries and local labour markets. Regional Science Association Annual Conference, Manchester, September

Budd L, Whimster S (eds) 1992 *Global finance and urban living.* Routledge, London

Buiges P, Ilzkovitz F, Lebrun J-F, Sapir A (eds) 1993 Market services and European integration: the challenges of the 1990s. *European Economy: Social Europe*, No. 3. Commission of the European Communities, Brussels

Burtenshaw D, Bateman M, Ashworth G J 1981 *The city in Western Europe.* Wiley, Chichester

Burton D 1990 Competition in the UK retail financial services sector: some implications for the spatial distribution and function of bank branches. *The Service Industries Journal* **10**: 571–88

Cabinet Office 1988 *Improving management in government: the next steps.* Efficiency Unit, HMSO, London

Carvey R 1988 American downtowns: past and present attempts at revitalization. *Built Environment* **14**: 46–70

Castells M 1977 *The urban question: a Marxist approach.* Edward Arnold, London

Castells M 1985 High technology, economic restructuring and the urban–regional process in the United States. In Castells M (ed) *High technology space and society.* Sage, Beverly Hills, CA

Castells M 1987 Technological change, economic restructuring and the spatial division of labor. In Muegge H, Stohr W, Hesp P, Stuckey B (eds) *International economic restructuring and the regional community.* Avebury, Aldershot, Hants: 45–63

Castells M 1989 *The informational city.* Basil Blackwell, Oxford

Cecchini P *et al* 1988 *The European challenge 1992: the benefits of a single market.* Wildwood House, Aldershot, Hants

Cheshire P, Hay D 1989 *Urban problems in Western Europe.* Unwin Hyman, London

Christaller W 1966 *Central places in Southern Germany.* Prentice Hall, Englewood Cliffs, NJ

Christopherson S 1989 Flexibility in the US economy and the emerging spatial division of labour. *Transactions of the Institute of British Geographers* **14**: 131–43

Christopherson S, Noyelle T 1992 The US path towards flexibility and productivity: the re-making of the US labour market in the 1980s. In Ernste H, Meier V (eds) *Regional development and contemporary industrial response*. Belhaven, London: 163–78

Clairmonte F, Cavanagh J 1984 Transnational corporations and services: the final frontier. *Trade and Development* **5**: 215–73

Clark C A 1940 *The conditions of economic progress*. Macmillan, London

Clutterbuck D (ed) 1985 *New patterns of work*. St Martin's Press, Gower, New York

Coffey W, Bailly A S 1992 Producer services and systems of flexible production. *Urban Studies* **29**: 857–68

Coffey W, Polese M 1987 Intrafirm trade in business services: implications for the location of office-based services. *Papers of the Regional Science Association* **62**: 71–80

Coffey W, Polese M 1992 Producer services and regional development: a policy-oriented perspective. *Papers of the Regional Science Association* **67**: 13–28

Cohen S, Zysman J 1987 *Manufacturing matters. The myth of the post-industrial society*. Basic Books, New York

Collier D 1983 The service sector revolution: the automation of services. *Long Range Planning* **16**: 10–20

Cooke P (ed) 1989 *Localities*. Unwin Hyman, London

Corby S 1991 Civil service decentralisation: reality of rhetoric? *Personnel Management* **February**: 39–42

Coriat B, Petit P 1991 Deindustrialisation and tertiarisation: towards a new economic regime. In Amin A and Dietrich M (eds) *Towards a new Europe. Structural change in the European economy*. Edward Elgar, Aldershot, Hants: 18–48

Cox J C, Warf B 1991 Wall Street layoffs and service sector incomes in the New York metropolitan region. *Urban Geography* **12**: 226–39

Crang P, Martin R L 1991 Mrs Thatcher's vision of the 'new Britain' and the other sides of the 'Cambridge phenomenon'. *Environment and Planning D: Society and Space* **9**: 91–116

Crompton R, Jones G 1984 *White collar proletariat: deskilling and gender in clerical work*. Macmillan, London

Crompton R, Sanderson K 1990 *Gendered jobs and social change.* Unwin Hyman, London

Crum R E, Gudgin G 1977 *Non production activities in UK manufacturing industry,* Regional Policy Series 3. Commission of the European Communities, Brussels

Curson C (ed) 1986 *Flexible patterns of work.* Institute of Personnel Management, London

Daly M T, Stimson R J 1994 Dependency in the modern global economy: Australia and the changing face of Asian finance. *Environment and Planning A* **26**: 415–34

Daniels P W 1975 *Office location: an urban and regional study.* Bell, London

Daniels P W (ed) 1979 *Spatial patterns of office growth and location.* Wiley, Chichester, Sussex

Daniels P W 1982 *Service industries: growth and location.* Cambridge University Press, Cambridge

Daniels P W 1985 *Service industries: a geographical appraisal.* Methuen, Andover, Hants

Daniels P W 1986a Producer services in the UK space economy. In Martin R, Rowthorn B (eds) *The geography of deindustrialisation.* Macmillan, London

Daniels P W 1986b Foreign banks and metropolitan development: a comparison of London and New York. *Tijdschrift voor Economische en Sociale Geografie* **77**, 269–87

Daniels P W (ed) 1991a *Services and metropolitan development, international perspectives.* Routledge, Chapman and Hall, Andover, Hants

Daniels P W 1991b A world of services? *Geoforum* **22**: 359–76

Daniels P W 1993 *Service industries in the world economy.* Blackwell, Oxford

Daniels P W, Moulaert F (eds) 1991 *The changing geography of advanced producer services. Theoretical and empirical perspectives.* Belhaven Press, London

Daniels P W, Thrift N 1987 *The geographies of the UK service sector: a survey.* Working paper on Producer Services No. **6**, Department of Geography, Portsmouth Polytechnic and University of Bristol

Daniels P W, Leyshon A, Thrift N J 1988 Trends in the growth and location of professional producer services: UK property consultants. *Tijdschrift voor Economische en Sociale Geografie* **79**: 162–74

Daniels P W, Van Dinteren J H J, Monnoyer M C 1992 Consultancy services and the urban hierarchy in W. Europe. *Environment and Planning A* **24**: 1731–48

Daniels P W, Illeris S, Bonamy J, Phillipe J (eds) 1993 *The geography of services*. Frank Cass, London

Davies R, Champion A (eds) 1983 *The future for the city centre*. Academic Press, London

Dawson J 1988 Futures for the high street. *The Geographical Journal* **154**: 13–16

Delaunay J C, Gadrey J 1992 *Services in economic thought. Three centuries of debate*. Kluwer, Boston, MA

Department of the Environment 1985 *Competition in the provision of local authority services*. HMSO, London

De Smidt M 1984 Office location and the urban functional mosaic; a comparative study of five cities in the Netherlands. *Tijdschrift voor Economische en Sociale Geografie* **75**: 110–21

De Smidt M 1985 Relocation of government services in the Netherlands. *Tijdschrift voor Economische en Sociale Geografie* **76**: 232–6

De Smidt M 1990 The 'informal' functional city-region: cooperation or competition? *Tijdschrift voor Economische en Sociale Geografie* **81**: 280–8

Dex S 1988 Gender and the labour market. In Gallie D (ed) *Employment in Britain*. Blackwell, Oxford: 281–309

Dicken P 1992 *Global shift*. Harper and Row, London

Dicken P, Lloyd P 1981 *Modern Western society: a geographical perspective on work health and wellbeing*. Harper & Row, London

Dicken P, Thrift N 1992 The organisation of production and the production of organisation. *Transactions of the Institute of British Geographers* **17**: 279–91

Drennan M P 1989 Information intensive industries in metropolitan areas of the USA. *Environment and Planning A* **21**: 1603–18

Drennan M P 1992 Gateway cities: the metropolitan sources of US producer service exports. *Urban Studies* **29**: 217–35

Drucker P 1988 The coming of the new organisation. *Harvard Business Review* **88**: 45–53

Ducatel K, Blomley N 1990 Rethinking retail capital. *International Journal of Urban and Regional Research* **14**: 207–27

Duncan S, Goodwin M 1988 *The local state and uneven development.* Polity Press, Cambridge

Dunning J H, Norman G 1987 The locational choice of offices of international companies. *Environment and Planning A* **19**: 613–31

Edwards L 1983 Towards a process model of office-location decision-making. *Environment and Planning A* **15**: 1327–42

Elfring T 1984 The main factors and underlying causes of the shift to services. *The Service Industries Journal* **9**: 337–56

Elfring T 1988 *Service employment in advanced economies.* Gower, Aldershot, Hants

Elfring T 1989 New evidence on the expansion of service employment in advanced economies. *Review of Income and Wealth* **35**: 409–40

Elkin T, McLaren D 1991 *Reviving the city: towards sustainable urban development.* Friends of the Earth, London

Enderwick P 1989 *Multinational service firms.* Routledge, Chapman and Hall, Andover, Hants

Ernste H, Jaeger C (eds) 1989 *Information society and spatial structure.* Belhaven, London

Evans A W 1973 The location of the headquarters of industrial companies. *Urban Studies* **10**: 387–95

Featherstone M 1991 *Consumer culture and postmodernism.* Sage, London

Feketekuty G, 1988, *International trade in services: an overview and blueprint for negotiations.* Ballinger, Cambridge, MA

Fik J, Malecki E J, Amey R 1993 Trouble in paradise? Employment trends and forecasts for a service-orientated economy. *Economic Development Quarterly* **7**: 358–72

Findlay A, Gould W T S 1989 Skilled international migration: a research agenda. *Area* **21**: 3–11

Fine B 1993 Modernity, urbanism and modern consumption: a comment. *Environment and Planning D: Society and Space* **11**: 599–601

Fine B, Leopold E 1993 *The world of consumption.* Routledge, Chapman and Hall, Andover, Hants

Fisher A G B 1935 *The clash of progress and security.* Macmillan, London

Fisher A G B 1939 Primary, secondary, tertiary production. *Economic Record.* June: 24–38

Foley D L 1956 Factors in the location of administrative offices. *Papers and Proceedings, Regional Science Association* **2**: 318–26

Forester T 1987 *High-tech society.* Blackwell, Oxford

Fothergill S, Gudgin G 1982 *Unequal growth; urban and regional employment change in the UK.* Heinemann, London

Freeman C (ed) 1983 *Long waves in the world economy.* Butterworth, London

Freeman C 1987 *Technology, policy and economic performance.* Frances Pinter, London

Friedmann J 1986 The world city hypothesis. *Development and Change.* **17**: 69–83

Friedmann J, Wolff G 1982 World city formation: an agenda for research and action. *International Journal for Urban and Regional Research* **6**: 309–44

Frobel F *et al.* 1980 *The new international division of labour.* Cambridge University Press, Cambridge

Frost M, Spence N 1993 Global city characteristics and central London's employment. *Urban Studies* **30**: 547–58

Fuchs V R 1968 *The service economy.* National Bureau for Economic Research, Columbia University Press, New York

Fuchs V R 1969 *Production and productivity in the service industries.* National Bureau for Economic Research, New York

Gad G H K 1979 Face-to-face linkages and office decentralisation: a study of Toronto. In Daniels P W (ed) *Spatial patterns of office growth and location.* Wiley, London: 277–324

247

Gentle C J 1993 *The financial services industry.* Avebury, Aldershot

Gentle C J, Marshall J N 1992 The deregulation of the financial services industry and the polarization of regional economic prosperity. *Regional Studies* **26**: 581–6

Gentle C J, Marshall J N 1993 Corporate restructuring in the financial services industry: converging markets and divergent organisational structures. *Geografiska Annaler B* **74**: 87–104

Gentle C J, Marshall J N, Coombs M G 1991 Business reorganization and regional development: The case of the British Building Societies Movement. *Environment and Planning A* **23**: 1759–77

Gershuny J 1978 *After industrial society: the emerging self-service economy?* Macmillan, London

Gershuny J 1985 *Social innovation.* Oxford University Press, Oxford

Gershuny J 1987 The future of service employment. In Giarini O (ed) *The emerging service economy.* Pergamon, Oxford: 105–24

Gershuny J, Miles I 1983 *The new service economy.* Frances Pinter, London

Gershuny J, Pahl R E 1979 Work outside employment: some preliminary speculations. *New Universities Quarterly* **34**: 120–35

Gertler M 1988 The limits to flexibility: comments on the post-Fordist vision of production and its geography. *Transactions of the Institute of British Geographers* **13**: 419–32

Gertler M 1992 Flexibility revisited: districts, nation-states and the forces of production. *Transactions of the Institute of British Geographers* **17**: 259–78

Gibbs D 1988 Restructuring in the Manchester clothing industry: technical change and the interrelationships between manufacturers and retailers. *Environment and Planning A* **20**: 1219–33

Gillespie A E, Green A 1987 The changing geography of producer service employment in Britain. *Regional Studies* **21**: 397–412

Gillespie A E, Robins K 1991 Non-universal service? Political economy and communications geography. In Brotchie J, Batty M, Hall P, Newton P (eds) *Cities of the 21st century. New technology and spatial systems.* Longman Cheshire, Harlow, Essex: 159–70

Gillespie A E, Williams H 1988 Telecommunications and the reconstruction of regional comparative advantage. *Environment and Planning A* **29**: 1311–21

Gillespie A E, Goddard J, Thwaites A T, Smith I, Robinson F 1984 *The effects of new information technology on the less-favoured regions of the European Communities,* Regional Policy Studies 23. Commission of the European Communities, Brussels

Gilmer R 1990 Identifying service sector exports from major Texas cities. *Federal Reserve Bank of Dallas, Economic Review* **July**: 1–16

Gilmer R, Keil S, Mack R 1989 The service sector in a hierarchy of rural places: potential for export activity. *Land Economics* **65**: 217–27

Glasmeier A, Borchard G 1989 From branch plants to back offices: prospects for rural services growth. *Environment and Planning A* **21**: 1565–84

Glennie P D, Thrift N J 1992 Modernity, urbanism and modern consumption. *Environment and Planning D: Society and Space* **10**: 423–42

Glennie P D, Thrift N J. 1993 Modern consumption: theorising commodities and consumers. *Environment and Planning D: Society and Space* **11**: 603–6

Goddard J 1973 Office linkages and location: a study of communications and spatial patterns in central London. *Progress and Planning* **1**: 109–232

Goddard J 1975 *Office location in urban and regional development.* Oxford University Press, Oxford

Goddard J 1979 Office location in urban and regional development. In Daniels P W (ed) *Spatial patterns of office growth and development.* Wiley, London: 37–62

Goddard J, Morris D 1976 The communications factor in office decentralisation. *Progress in Planning* **6**: 1–80

Goddard J, Pye R 1977 Telecommunications and office location. *Regional Studies* **11**: 19–30

Goe W R 1990 Producer services, trade and the social division of labour. *Regional Studies* **24**: 327–42

Goldberg M A, Helsley R W, Levi M D 1989 The location of international financial activity: an interregional analysis. *Regional Studies* **23**: 1–7

Goldberg M A, Helsley R W, Levi M D 1991 The growth of international financial services and the evolution of international financial centres: a regional and urban economic approach. In Daniels P W (ed) *Services and metropolitan development: international perspectives.* Routledge, Chapman and Hall, London: 44–65

Goodwin M, Duncan S, Halford S 1993 Regulation theory, the local state and the transition of urban politics. *Environment and Planning D: Society and Space* **11**: 67–88

Gorz A 1985 *Paths to paradise: on the liberation from work.* Pluto, London

Goss J 1992 Modernity and post-modernity in the retail landscape. In Anderson K, Gale F (eds) *Inventing places: studies in cultural geography.* Wiley, New York: 158–77

Gottmann J 1961 *Megalopolis: the urbanized northeastern seaboard of the United States.* Twentieth Century Fund, New York

Gottmann J 1974 The dynamics of large cities. *The Geographical Journal* **140**: 254–61

Gottmann J 1976 Metropolitan systems around the world. *Ekistics* **41**: 109–13

Gottmann J 1977 The role of capital cities. *Ekistics* **44**: 240–3

Gottmann J 1978 Urbanisation and employment: towards a general theory. *Town Planning Review* **49**: 393–401

Gottmann J 1979 Office work and the evolution of cities. *Ekistics* **46**: 4–7

Gottmann J 1983 *The coming of the transactional city.* Institute for Urban Studies, University of Maryland, College Park, MD

Gottmann J 1989 What are cities becoming the center of? In Knight R, Gappert G (eds) *Cities in a global society.* Sage, Newbury Park, CA: 58–67

Gough I 1972 Marx's theory of productive and unproductive labour. *New Left Review* **76**: 47–72

Gough I 1979 *The political economy of the Welfare State,* Macmillan, London

Graham N, Beatson M, Wells W 1989 1977 to 1987: a decade of service. *Employment Gazette* **99**: January, 45–50

Green A, Howells J 1988 Information services and spatial development in the UK economy. *Tijdschrift voor Economische en Sociale Geografie* **79**: 266–77

Greenfield H 1966 *Manpower and the growth of producer services.* Columbia University Press, New York

Griffiths M 1994 The economic impact of Lightwater Valley. Undergraduate dissertation, Department of Geography, University of Newcastle

Grubel H G 1988 Direct and embodied trade in services or where is the service trade problem? In Lee C H, Naja S (eds) *Trade and investment in services in the Asia–Pacific Region.* Westview Press, Boulder, CO: 53–72

Guardian 1993 HQ cuts in IBM shake-up. **1 July:** 15

Gudgin G, Crum R, Bailey S 1979 White collar employment in UK manufacturing. In Daniels P W (ed) *Spatial patterns of office growth and location* Wiley, Chichester: 127–57

Hakim C 1987 Trends in the flexible workforce. *Employment Gazette.* **95:** 549–60

Hall P 1984 *The world cities* 3rd Edn. Weidenfeld and Nicolson, London

Hall P 1987 The anatomy of job creation: nations, regions and cities in the 1960s and 1970s. *Regional Studies* **21:** 95–106

Hall P 1988 The geography of the fifth Kondratiev. In Allen J, Massey D (eds) *Uneven re-development: cities and regions in transition.* Hodder & Stoughton, London: 51–67

Hall P 1989 The rise and fall of great cities: economic forces and population responses. In Lawton R (ed) *The rise and fall of great cities.* Belhaven Press, London: 20–31

Handy C 1985 *The future of work.* Blackwell, Oxford

Hansen N 1990 Do producer services induce regional economic development? *Journal of Regional Science* **30:** 465–76

Harrington J W, MacPherson A D, Lombard J R 1991 Interregional trade in producer services: review and synthesis. *Growth and Change* **22:** 75–94

Harris L 1988 The UK economy at the crossroads. In Allen J, Massey D (eds) *The economy in question.* Sage, London: 7–44

Harris R I D 1987 The role of manufacturing in regional growth. *Regional Studies* **21:** 301–12

Harvey D 1985 *The urbanisation of capital.* Blackwell, Oxford

Harvey D 1989a *The condition of postmodernity*. Blackwell, Oxford

Harvey D 1989b From managerialism to entrepreneurialism: the transformation in urban governance in late capitalism. *Geografiska Annaler B* **71**: 3–18

Heald D 1983 *Public expenditure*. Martin Robertson, Oxford

Healey M 1994 Teaching economic geography in UK higher education institutions. *Journal of Geography in Higher Education* 18: 70–79

Healey M, Ilbery BC 1990 *Location and change, perspectives on economic geography*. Oxford University Press, Oxford

Hennessey P 1989 *Whitehall*. Secker and Warburg, London

Hepworth M 1986 The geography of technical change in the information economy. *Regional Studies* **20**: 407–24

Hepworth M 1987 Information technology as spatial systems. *Progress in Human Geography* **11**: 157–80

Hepworth M 1989 *Geography of the information economy*. Belhaven Press, London

Hessels M 1992 *Locational dynamics of business services; an intrametropolitan study of the Randstad Holland*. Nederlandse Geografische Studies, the Royal Dutch Geographical Society/Faculty of Geographical Sciences Utrecht University

Hill T P 1977 On goods and services. *Review of Income and Wealth* **23**: 315–38

Holti R, Stern E 1986 *Distance working*. FAST, Commission of the European Communications, Brussels

Howells J 1988 *Economic, technological and locational trends in European services*. Avebury, Aldershot, Hants

Howells J 1989 Externalisation and the formation of new industrial operations: a neglected dimension in the dynamics of industrial location. *Area* **21**: 289–99

Howells J, Green A 1986 Location, technology and industrial organisation in UK services. *Progress in Planning* **27**: 83–184

Hoyle B S, Hilling D (eds) 1984 *Seaport systems and spatial change*. Wiley, Chichester

Hudson R 1988 *Wrecking a region: state policies, party politics and regional change in North East England.* Pion, London

Hutton T, Ley D 1987 Location, linkages and labor: the downtown complex of corporate activities in a medium size city, Vancouver, British Columbia. *Economic Geography* 6: 126–41

Illeris S 1989a Formal employment and informal work in household services. *The Service Industries Journal* 13: 94–109

Illeris S 1989b *Services and regions in Europe.* Avebury, Aldershot, Hants

Inman R (ed) 1985 *Managing the service economy: prospects and problems.* Cambridge University Press, Cambridge

Jaeger C, Durrenberger G 1991 Services and counterurbanisation: the case of central Europe. In Daniels P W (ed) *Services and metropolitan development. International perspectives.* Routledge, London: 107–28

James V Z, Marshall J N, Walters N 1979 *Telecommunications and office location,* Final Report to DOE. CURDS, University of Newcastle upon Tyne

Jefferson C W, Trainor M 1993 Public sector employment in regional development. *Environment and Planning A* 25: 1319–38

Jensen J 1989 The talents of women, the skills of men: flexible specialisation and women. In Wood S (ed) *The transformation of work?* Unwin Hyman, London: 141–55

Jessop B 1991a The welfare state in the transition from Fordim to post-Fordism. In Jessop B, Kastendiek H, Neilsen K, Pedersen O (eds) *The politics of flexibility. Restructuring state and industry in Britain, Germany and Scandinavia.* Edward Elgar, Aldershot, Hants: 82–105

Jessop B 1991b Thatcherism and flexibility: the white heat of a post Fordist revolution. In Jessop B, Kastendiek H, Neilsen K, Pedersen O (eds) *The politics of flexibility. Restructuring state and industry in Britain, Germany and Scandinavia.* Edward Elgar, Aldershot, Hants: 135–61

Jessop B, Kastendiek H, Neilsen K, Pedersen O (eds) 1991 *The politics of flexibility. Restructuring state and industry in Britain, Germany and Scandinavia.* Edward Elgar, Aldershot, Hants.

Johnson C 1991 Financial services to slow down. *Lloyds Bank Economic Bulletin,* No. 145, January

Johnson J H, Salt J (eds) 1990 *Labour migration. The internal geographical mobility of labour in the developed world.* Fulton, London

Johnson P S, Thomas R B 1990 Measuring the local employment impact of a tourist attraction: an empirical study. *Regional Studies* **24**: 395–403

Johnson P S, Thomas R B 1992 *Tourism, museums and the local economy*. Edward Elgar, Aldershot, Hants

Jones P 1989 The high street fights back. *Town and Country Planning* **58**: 43–5

Julius D 1990 *Global companies and public policy*. Frances Pinter, London

Keeble D 1990 Small firms, new firms and uneven regional development. *Area* **22**: 234–5

Keeble D, Bryson J, Wood P A 1991 Small firms, business service growth and regional development in the United Kingdom. *Regional Studies* **25**: 439–57

Keil S R, Mack R S 1986 Identifying export potential in the service sector. *Growth and Change* **17**: 1–10

Kendrick J W 1985 Measurement of output and productivity in the service sector. In Inman R P (ed) *Managing the service economy: prospects and problems*. Cambridge University Press, Cambridge: 111–31

Kirn T J 1987 Growth and change in the service sector of the US: a spatial perspective. *Annals of the Association of American Geographers* **77**: 353–72

Kumar K 1978 *Prophecy and progress: the sociology of industrial and post-industrial society*. Penguin, Harmondsworth

Kutscher R E R 1988 Growth of services employment in the United States. In Guile B, Quinn JB (eds) *Technology in services: policies for growth, trade and employment*. National Academic Press, Washington, DC

Ladd, H F 1992 Population growth, density and the costs of providing public services. *Urban Studies* **29**: 273–95

Lambooy J G, Tordoir P P 1985 Professional services and regional development: a conceptual approach; Paper presented on Services and Regional Development, FAST, Commission of the European Communities, Brussels

Lash S, Urry J 1987 *The end of organised capitalism*. Polity, Cambridge

Law C 1992 Urban tourism and its contribution to economic regeneration. *Urban Studies* **29**: 599–618

Law C 1994 *Urban tourism: attracting visitors to large cities.* Mansell Publishing, London

Lea J 1988 *Tourism and development in the Third World.* Routledge, Chapman and Hall, Andover, Hants

Lee C H 1986 *The British economy since 1700: a macro economic perspective.* Cambridge University Press, Cambridge

Lee C H, Naja S 1988 *Trade and investment in services in the Asia–Pacific Region.* Westview Press, Boulder, Co

Leidner R 1991 Serving hamburgers and selling insurance: gender, work, and identity in interactive service jobs. *Gender and Society* **5**: 154–77

Leo P-Y, Phillippe J 1991 Networked producer services: local markets and global development. In Daniels P W (ed) *Services and metropolitan development.* Routledge, Chapman & Hall, Andover, Hants: 305–24

Lettre de Liason des Service 1990 Editorial, September, Centre d'echange et d'information sur les activities de service. CEDEX, Lyon

Leveson I 1980 *Productivity in services: issues for analysis.* Hudson Institute, New York

Leyshon A, Thrift N 1989 South goes North? The rise of British provincial financial centres. In Lewis J, Townsend A (eds) *The North–South divide.* Paul Chapman, London: 114–56

Leyshon A, Thrift N 1990 *The power of money: global financial capital and the restructuring of the international financial system.* Department of Geography and Earth Resources, University of Hull

Leyshon A, Thrift N 1992 Liberalisation and consolidation: the Single European Market and the remaking of European financial capital. *Environment and Planning A* **24**: 49–81

Leyshon A, Thrift N 1993 The restructuring of the UK financial services industry in the 1990s: a reversal of fortune? *Journal of Rural Studies* **3**: 223–41

Leyshon A, Daniels P W, Thrift N 1988 Large accounting firms in the UK and spatial development. *The Service Industries Journal* **8**: 317–46

Lipietz A 1986 New tendencies in the international division of labour: regimes of accumulation and modes of regulation. In Scott A, Storper M (eds) *Production work, territory.* Allen & Unwin, Winchester MA: 16–40

Lipietz A 1993 The local and the global: regional individuality or interregionalism. *Transactions of the Institute of British Geographers* **18**: 8–18

Little J, Peake L, Ritchardson P (eds) 1988 *Women in cities*. Macmillan, London: 1–20

Lovering J 1985 Regional economic development and the role of defence procurement: a case study. *Built Environment* **11**: 193–206

Lovering J 1989 The restructuring debate. In Peet R and Thrift N (eds) *New models in geography*, Vol. 1. Unwin Hyman, London: 198–223

Lovering J 1990 Fordism's unknown successor: a comment on Scott's theory of flexible accumulation and the re-emergence of regional economies. *International Journal of Urban and Regional Research* **14**: 159–74

Lovering J, Boddy M 1988 The geography of military industry in Britain. *Area* 41–51

Lowe M S 1991a Lollipop jobs for pin money? *Area* **23**: 344–7

Lowe M S 1991b Trading places: retailing and local economic development at Merry Hill, West Midlands. *East Midland Geographer* **14**: 31–48

Lowe M S, Gregson N 1989 Nannies, cooks, cleaners, au pairs: new issues for feminist geography. *Area* **21**: 415–17

McDowell L 1983 Towards an understanding of the gender division of urban space. *Environment and Planning D: Society and Space* **1**: 59–72

McDowell L 1989 Gender divisions. In Hamnett C, McDowell L, Sarre P (eds) *The changing social structure*. Sage, Open University Press, Milton Keynes: 158–98

McDowell L 1991 Life without Father and Ford: the new gender order of post-Fordism. *Transactions of the Institute of British Geographers* **16**: 400–19

McDowell L 1992 Doing gender: feminism, feminists and research methods in human geography. *Transactions of the Institute of British Geographers* **17**: 399–416

McDowell L 1993 Space, place and gender relations: Part 1. Feminist empiricism and the geography of social relations. *Progress in Human Geography* **17**: 157–79

McGregor A, Sproull A 1992 Employees and the flexible workforce. *Employment Gazette* **100**: 225–34

McKinnon A 1990 The advantages and disadvantages of centralised distribution in Fernie J (ed) *Retail distribution management*. Kogan Page, London: 75–89

MacPherson A 1988 New product development among small Toronto manufacturers: empirical evidence on the role of technical service linkages. *Economic Geography* **64**: 62–75

MacPherson A 1991 Interfirm information linkages in an economically disadvantaged region: an empirical perspective from metropolitan Buffalo. *Environment and Planning A* **23**: 59–66

Magdoff H, Sweezy P 1987 *Stagnation and the financial explosion*. Monthly Review Press, New York

Malecki E J 1984 Military spending and the US defense industry: regional patterns of contracts and subcontracts. *Environment and Planning C: Government and Policy*. **2**: 31–44

Malecki E J 1991 *Technology and economic development: the dynamics of local regional and national change*. Longman, Harlow, Essex

Mandel E 1975 *Late capitalism*. New Left Books, London

Mansfield Y 1990 Spatial patterns of international tourist flows: towards a theoretical framework. *Progress in Human Geography* **14**: 372–90

Marquand J 1979 *The service sector and regional policy in the UK*. Research Series 29, Centre for Environmental Studies Ltd, London

Marquand J 1983 The changing distribution of service employment. In Goddard J B, Champion A G (eds) *The urban and regional transformation of Britain*. Methuen, London

Marshall J N 1978 *Business travel and regional office employment* Discussion Paper No. 6., CURDS, University of Newcastle upon Tyne

Marshall J N 1979 Corporate organisation and regional office employment. *Environment and Planning A* **1**: 553–63

Marshall J N 1982 Linkages between manufacturing industry and business services. *Environment and Planning A* **14**: 1523–40

Marshall J N 1983 Business service activities in British provincial conurbations. *Environment and Planning A* **15**: 1343–59

Marshall J N 1984 Information technology changes corporate office activity. *GeoJournal* **92**: 171–8

Marshall J N 1985 Business services, the regions and regional policy. *Regional Studies* **19**: 353–63

Marshall J N 1989 Corporate reorganisation and the geography of services: evidence from the motor vehicle aftermarket in the West Midlands Region of the UK. *Regional Studies* **23**: 139–50

Marshall J N 1990 Reorganising the British civil service: how are the regions being served? *Area* **22**: 246–55

Marshall J N 1992 The growth of service activities and the evolution of spatial disparities. In Townroe P, Martin R (eds) *Regional development in the 1990s*. Jessica Kingsley, London: 204–13

Marshall J N forthcoming Services and industrial policy. In Coates D (ed) *Industrial policy in Britain*. Macmillan, London

Marshall J N, Alderman N 1991 Rolling back the frontiers of the state: Civil service reorganisation and relocation in Britain. *Growth and Change* **22**: 51–74

Marshall J N, Alderman N, Thwaiter, A 1991 Civil Service relocation and the English regions. *Regional Studies* **25**: 499–510

Marshall J N, Jaeger C 1990 Service activities and uneven spatial development in Britain and its European partners: deterministic fallacies and new options. *Environment and Planning A* **22**: 1337–54

Marshall J N, Raybould S 1993 New corporate structures and the evolving geography of white collar work. *Tijdschrift voor Economische en Sociale Geografie* **84**: 362–76

Marshall J N, Wood P A 1992 The role of services in urban and regional development: recent debates and new directions. *Environment and Planning A* **24**: 1255–70

Marshall J N, Damesick P, Wood P A 1987 Understanding the location and role of producer services in the United Kingdom. *Environment and Planning A* **19**: 575–96

Marshall J N *et al.* 1988 *Services and uneven development*. Oxford University Press, Oxford.

Marshall J N, Gentle C J, Raybould S, Coombes M 1992 Regulatory change, corporate restructuring and the spatial development of the financial sector. *Regional Studies* **26**: 453–68

Martin P 1990 Return of the invisible hand. *Financial Times*, 16th May

Martinelli F 1991 Producer services, location and regional development. In Daniels P W (ed) *The changing geography of advanced producer services*. Belhaven, London: 70–90

Massey D 1979a A critical evaluation of industrial location theory. In F E Hamilton, G D R Linge (eds) *Spatial analysis and the industrial environment*. Wiley, Chichester, Sussex: 57–72

Massey D 1979b In what sense a regional problem? *Regional Studies* **13**: 233–43

Massey D 1984 *Spatial divisions of labour*. Macmillan, London

Massey D 1991 Flexible sexism. *Environment and Planning D: Society and Space* **9**: 31–57

Massey D, Meegan R 1982 *The anatomy of job loss*. Methuen, Andover, Hants

Mather G 1992 Interview quoted in Phillips M, The tenders trap of Civil Service Plc. *The Guardian* **21 July**: 21

Mathieson A, Wall G 1982 *Tourism: economic, physical and social impacts*. Longman, Harlow, Essex

Merrifield A 1993 The struggle over place: redeveloping American Can in Southeast Baltimore. *Transactions of the Institute of British Geographers* **18**: 102–21

Michalak W Z, Fairbairn K J 1993 The producer service complex of Edmonton: the role and organization of producer service firms in a peripheral economy. *Environment and Planning A* **25**: 761–78

Miles I 1988 *Home informatics: information technology and the transformation of everyday life*. Frances Pinter, London

Miles I 1993 Services in the new industrial economy. *Futures* **July/August**: 653–72

Miles I, Rush H, Turner K, Bessant J 1989 *Information horizons*. Edward Elgar, Aldershot, Hants

Miller D 1987 *Material culture and mass consumption*. Blackwell, Oxford

Mills S 1989 Tourism and leisure-setting the scene. *Tourism Today* **6**: 18–21

Mishra R 1984 *The Welfare State in Crisis*. Harvester, Chichester

Mohan J 1988a Spatial aspects of health-care employment in Britain: 1 aggregate trends. *Environment and Planning A* **20**: 7–23

Mohan J 1988b Restructuring, privatization and the geography of health care provision in England, 1983–87. *Transactions of the Institute of British Geographers* **13**: 449–65

Mohan J 1991 The internationalisation and commercialisation of health-care in Britain. *Environment and Planning A* **23**: 853–67

Mohan J 1992 Public expenditure, public employment and the regions. In Townroe P, Martin R (eds) *Regional development in the 1990s. The British Isles in transition.* Jessica Kingsley, London: 222–8

Mohan J, Lee R 1989 Unbalanced Growth? Public services and labour shortages in a European core region. In Breheny M, Congdon P (eds) *Growth and change in a core region.* Pion, London: 33–54

Molyneux P 1989 1992 and its impact on local and regional banking markets. *Regional Studies* **23**: 523–34

Momigliano F, Siniscalco O 1983 The growth of service employment: A reappraisal. *Banca Nazionale del Lavoro, Quarterly Review* **142**: 269–306

Moore G. Parnell R 1986 Private sector involvement in local authority service delivery. *Regional Studies* **20**: 253–7

Morris J L 1988 Producer services and the regions: the case of large accountancy firms. *Environment and Planning A* **20**: 741–59

Morris J L 1987 Industrial restructuring, foreign direct investment and uneven development: the case of Wales. *Environment and Planning* **19**: 205–24

Moss M L 1987a Telecommunications, world cities and urban policy. *Urban Studies* **24**: 534–46

Moss M L 1987b Telecommunications: shaping the future. Paper presented at the conference on America's New Economic Geography, April, Washington, DC

Moss M L 1987c Telecommunications and international financial centers. In Brotchie J, Hall P, Newton P (eds) *The spatial impact of technological change.* 75–88, Croom Helm, London: 75–88

Moss M L, Brion J G 1991 Foreign banks, telecommunications, and the central city. In Daniels P W (ed) *Services and metropolitan development: international perspectives.* Routledge, Chapman & Hall, London: 265–84

Moulaert F, Gallouj C 1993 The locational geography of advanced producer service firms: the limits of economies of agglomeration. *The Service Industries Journal* **13**: 91–106

Muligan G J 1991 *Communications and control*. Polity Press, Cambridge

Murphy P E 1985 *Tourism: a community approach*. Methuen, London

Myerscough J 1988 *The economic significance of the arts in Britain*. Policy Studies Institute, London

National Economic Development Office 1985 *IT Futures*. NEDO, London

National Economic Development Council 1988 *The future of the high street*. HMSO, London

Nelson K 1986 Labour demand, labour supply and the suburbanisation of low-wage office work. In Scott A, Storper M (eds) *Production, work, territory*. Allen & Unwin, Winchester, MA: 149–71

Nicolaides P 1989 *Liberalizing service trade*. Royal Institute of International Affairs, London

Noam E 1992 *Telecommunications in Europe* Oxford University Press, Oxford

Northern Region Strategy Team 1976 *Office activity in the northern region*, Technical Report **8**. HMSO, London

Noyelle T J 1983 The rise of advanced services. *American Planning Association Journal* **49**: 280–90

Noyelle T 1984 Rethinking public policy for the service era. *Economic Development Commentary* **8**: 12–17

Noyelle T J 1986 *Beyond industrial dualism: market and job segmentation in the new economy*. Westview Press, Boulder, CO

Noyelle T J, Peace P 1991 Information industries: New York's export base. In Daniels P W (ed) *Services and metropolitan development: international perspectives*, Routledge, Chapman & Hall, London: 285–304

Noyelle T J, Stanback T M 1984 *The economic transformation of American cities*. Rowman & Allanheld, Totowa, NJ

Nusbaumer J 1987 *Services in the Global Market*, Kluwer-Nijhoff, Boston MA

Ochel W, Wegner M 1987 *Service economies in Europe: opportunities for growth*. Pinter/Westview, London

OECD 1983 Part-time employment in OECD countries. *Employment Outlook* September: 43–67

OCED 1986 *Trends in the information economy, organisation for economic co-operation and development*. OECD, Paris

OCED 1992 *Labour Force Statistics 1970–90*. OECD, Paris

O'Connor K 1987 The location of services involved with international trade. *Environment and Planning A* **19**: 687–700

O'Farrell P N, Hitchens D M 1990a Research policy and review 32: producer services and regional development: a review of some major conceptual and policy issues. *Environment and Planning A* **22**: 1141–54

O'Farrell P N, Hitchens D M W N 1990b Producer services and regional development: key conceptual issues of taxonomy and quality measurement. *Regional Studies* **24**: 163–71

O'Farrell P N, Hitchens D M W N 1992 The competitiveness of business service firms: a matched comparison between Scotland and the South East of England. *Regional Studies* **26**: 519–35

O'Farrell P N, Moffat L A R, Hitchens D M W N 1993 Manufacturing demand for business services in a core and peripheral region: does flexible production imply vertical disintegration of business services? *Regional Studies* **27**: 385–400

O'Farrell P N, Wood P A, Moffat, L A R 1995 Export behaviour of business services, *Environment and Planning A* (forthcoming)

Office of Technology Assessment 1987 *International competition in services*. Washington DC

O'hUallachain B 1989 Agglomeration of services in American metropolitan areas. *Growth and Change* **20**: 34–49

O'hUallachain B 1991 The location and growth of business and professional services in American metropolitan areas, 1976–1986. *Annals of the Association of American Geographers* **81**: 254–70

Osborne D, Gaebler T 1992 *Reinventing government*. Addison Wesley

Pacini M *et al.* 1993 *La capitale recticolare*. Edizioni della Fondazione Giovanni Agnelli, Turin

Pahl R E (ed) 1984 *Divisions of labour*. Blackwell, Oxford

Pahl R E 1988a Some remarks on informal work, social polarization and the social structure. *International Journal for Urban and Regional Research* **12**: 247–67

Pahl R E (ed) 1988b *On work: historical, comparative and theoretical approaches*. Blackwell, Oxford

Painter J 1991 Regulation theory and local government. *Local Government Studies* **17**: 23–44

Parsons G 1972 The giant manufacturing companies and balanced regional growth in Britain. *Area* **4**: 99–103

Patton W, Markusen A 1991 The perils of overstating service sector growth potential: a study of linkages in distributive services. *Economic Development Quarterly* **5**: 197–212

Pearce D 1987 *Tourism today: a geographical analysis*. Longman, London

Peck J A, Tickell A 1991 Regulation theory and the geographies of flexible accumulation: transitions in capitalism: transitions in theory. *Spatial Policy Analysis WP 12*. School of Geography, University of Manchester

Peet R, Thrift N (eds) 1989 *New models in geography*, Vol. 1. Unwin Hyman, London

Perry A 1986 A theme for tourism. *The Geographical Magazine* **58**: 2–3

Perry M 1990a Business service specialisation and regional economic change. *Regional Studies* **24**: 195–210

Perry M 1990b The internationalisation of advertising. *Geoforum* **21**: 35–50

Perry M 1991 The capacity of producer services to generate regional growth: some evidence from a peripheral metropolitan economy. *Environment and Planning A* **23**: 1331–48

Persky J, Sclar E, Wiewel W 1991 *Does America need cities? An urban innovation strategy for national prosperity*. Economic Policy Institute, Washington, DC

Petit P 1986 *Slow growth and the service economy*. Frances Pinter, London

Phillips A, Taylor B 1986 Sex and skills. In Feminist Review (ed) *Waged work: a reader*. Virago, London: 54–66

Phillips M 1992 The tenders trap of Civil Service Plc. The *Guardian* **21 July**: 21

Pinch S P 1985 *Cities and services: the geography of collective consumption.* Routledge & Kegan Paul, London

Pinch S P 1989 The restructuring thesis and the study of public services. *Environment and Planning A* **21**: 905–26

Pinch S P 1993 Social polarisation in Britain and the US. *Environment and Planning A* **25**: 779–95

Pinch S P, Storey A 1992a Flexibility, gender and part-time work: evidence from a survey of the economically active. *Transactions of the Institute of British Geographers* **17**: 198–214

Pinch S P, Storey A 1992b Who does what, where? A household survey of the division of domestic labour in Southampton. *Area* **24**: 5–12

Pinch S P, Storey A 1992c Labour-market dualism: evidence from a survey of households in the Southampton city-region. *Environment and Planning A.* **24**: 571–89

Pinch S P, Mason C, Witt S 1989 Labour flexibility and industrial restructuring in the UK 'Sunbelt': the case of Southampton. *Transactions of the Institute of British Geographers* **14**: 418–34

Polese M 1982 Regional demand for business services and interregional service flows in a small Canadian region. *Papers of the Regional Science Association* **50**: 151–63

Pollert A 1988 Dismantling flexibility. *Capital and Class* **34**: 42–75

Porat M 1977 *The information economy: definition and measurement* SP-77-12(1). Office of Telecommunications, US Department of Commerce, Washington, DC

Porterfield S L, Pulver G C 1991 Exports, impacts and locations of service producers. *International Regional Science Review* **14**: 41–60

Portes A, Castells M, Benton L A 1989 *The informal economy: studies in advanced and less developed countries.* John Hopkins University Press, Baltimore, MD

Postner H H, Wesa L 1984 *Canadian productivity growth: an alternative (input output) analysis.* Economic Council of Canada, Ottawa

Prais S 1981 *The evolution of giant firms in Britain.* Cambridge University Press, Cambridge

Pratt G 1990 Feminist analyses of the restructuring of urban life. *Urban Geography* **11**: 594–605

Pred A R 1967 *Behaviour and location: foundations for a geographic and dynamic location theory, part 1, Lund Studies in Geography 27.* Lund.

Pred A 1977 *City systems in advanced economies.* Hutchinson, London

Pryke M 1991 An international city going global: spatial change in the City of London. *Environment and Planning D: Society and Space* **9**: 197–222

Pryke M 1994 Looking back on the space of a boom. *Environment and Planning A* **26**: 234–64

Rajan A 1984 *New technology and employment in insurance, banking and building societies: recent experience and future impact.* Gower, Aldershot, Hants

Rajan A 1985 Office technology and clerical skills. *Futures* **17**: 411–13

Rajan A 1987a *Services – The second industrial revolution?* Butterworths, London

Rajan A 1987b New technology and career progression in financial institutions. *The Service Industries Journal* **7**: 35–40

Rajan A, Cooke G 1986 The impact of information technology on employment in the financial services industry. *National Westminster Bank Review* **August**: 21–35

Rajan A, Fryatt J 1988 *Create or abdicate: the city's human resource choice for the '90s.* Witherby/Institute of Manpower Studies, University of Sussex

Redclift N, Mingione E (ed) 1988 *Beyond employment. Household, gender and Subsistence.* Blackwell, Oxford

Richardson R 1993 *Back officing front office functions: new telemediated services and economic development.* PICT Discussion Paper No. 10. CURDS, University of Newcastle upon Tyne

Riddle D 1986 *Service-led growth: the role of the service sector in world development.* Praeger, New York

Ridley N 1988 *The local right.* Centre for Policy Studies, London

Rimmer P J 1991 The global intelligence corps and world cities: engineering consultancies on the move. In Daniels P W (ed) *Services*

and metropolitan development: international perspectives. Routledge, Chapman & Hall, London: 66–106

Robertson K A 1983 Downtown retail activity in large American cities. *The Geographical Review* **73**: 314–23

Robertson J 1985 *Future work: jobs, self-employment and leisure after the industrial age*. Gower, Aldershot, Hants

Robertson J A S, Briggs J M, Goodchild A 1982 *Structure and employment prospects of the service industries*, Research Paper **3**. Department of Employment, London

Robins K, Gillespie A 1988 *Beyond Fordism, place, space and hyperspace*. PICT Discussion Paper No. **9** CURDS, University of Newcastle upon Tyne

Robinson O 1985 The changing labour market: The phenomenon of part-time employment in Britain. *National Westminster Bank Quarterly Review* 19–29

Robinson O, Wallace J 1984 Growth and utilisation of part-time labour in Great Britain. *Employment Gazette*, September: 391–7

Robson B T 1988 *Those inner cities: reconciling the social and economic aims of urban policy*. Clarendon, London

Rowthorn B 1986 De-industrialisation in Britain. In Martin R and Rowthorn B (eds) *The geography of de-industrialisation*. Macmillan, London: 1–30

Salt J 1984 High level manpower movements in northwest Europe and the role of careers: an explanatory framework. *International Migration Review* **17**: 633–52

Salt J 1992 Migration processes among the highly skilled in Europe. *International Migration Review* **26**: 484–505

Santos M 1979 Interdependencies in the urban economy. In Walman S (ed) *Ethnicity at work*. Macmillan, London: 214–26

Sapir A 1993 Sectoral dimension. In Buiges P *et al.* (eds) *Market services and European integration: the challenges of the 1990s*. European Economy: Social Europe, No. 3, Commission of the European Communities, Brussels: 23–40

Sassen S 1989 Finance and business services in New York City: international linkages and domestic effects. In Rodwin L, Sazanami H (eds) *Deindustrialization and regional economic transformation.* Unwin Hyman, Boston, MA: 132–54

Sassen S 1991 *The global city: New York, London, Tokyo.* Princeton University Press, Princeton, NJ

Sassen-Koob S 1984 The new labor demand in global cities. In Smith M P (ed) *Cities in transformation, class, capital and the state.* Sage, Beverley Hills, CA: 139–72

Savage M, Dickens P, Fielding T 1988 Some social and political implications of the contemporary fragmentation of the 'service class' in Britain. *International Journal of Urban and Regional Research* 12: 455–76

Savage M, Barlow J, Dickens P, Fielding A 1992 *Property, bureaucracy and culture: middle class formation in contemporary Britain.* Routledge, London

Savitch H V, Collins D, Saunders D, Markham J 1993 Ties that bind: Central cites, suburbs and the metropolitan region. *Economic Development Quarterly* 7: 341–57

Sayer A 1985 Industry and space: a sympathetic critique of radical research. *Environment and Planning D: Society and Space* 3: 3–29

Sayer A 1989 Post-Fordism in question. *International Journal of Urban and Regional Research* 13: 666–95

Sayer A, Walker R 1992 *The new social economy: reworking the division of labour.* Blackwell, Oxford

Scarpaci J L (ed) 1988 *Health services privatisation in industrial societies.* Rutgers University Press View, Brunswick

Schiller R 1988 Retail decentralisation – a property view. *The Geographical Journal* 154: 17–19

Schoenberger E 1987 Technological and organisational changes in automobile production: spatial implications. *Regional Studies* 21: 199–214

Schoenberger E 1989 Thinking about flexibility: a response to Gertler. *Transactions of the Institute of British Geographers* 14: 98–108

Schwartz A 1992 Corporate service linkages in large metropolitan areas. *Urban Affairs Quarterly* 28: 276–96

Scott A 1988a *New industrial spaces.* Pion, London

Scott A 1988b Flexible production systems and regional development: the rise of new industrial spaces in North America and Western Europe. *International Journal of Urban and Regional Research* **12**: 171–86

Scott A 1988c *Metropolis. From division of labor to urban form.* University of California Press, Berkley

Scott A 1992 The role of large producers in industrial districts: a case study of high technology systems houses in Southern California. *Regional Studies* **26**: 265–76

Scott A, Storper M (eds) 1986 *Production, work, territory.* Allen & Unwin, Winchester, MA

Segal Quince and Partners 1985 *The Cambridge phenomenon: the growth of high technology in a University town.* SQ and P. Mount Pleasant House, Cambridge

Semple R K 1973 Recent trends in the spatial concentration of corporate headquarters. *Economic Geography* **49**: 309–18

Semple R K, Phipps A G 1982 The spatial evolution of corporate headquarters within an urban system. *Urban Geography* **3**: 258–79

Shelp R K, Stephenson J C, Truitt N S, Wasow B 1984 *Service industries and economic development.* Praeger, New York

Shutt J, Whittington R 1987 Fragmentation strategies and the rise of small units. *Regional Studies* **21:** 13–24

Sinclair M T, Page S J 1993 The Euroregion: a new framework for tourism, and regional development. *Regional Studies* **27**: 475–83

Singelmann J 1978 *From agriculture to services: the transformation of industrial employment.* Sage, Beverley Hills

Sjoholt P 1993 The dynamics of services as an agent of regional change and development: the case of Scandinavia. In Daniels P W, Illeris S, Bonamy J, Philippe J (eds) *The geography of services.* Frank Cass, London: 36–50

Smith D M 1976 *Human geography: a welfare approach.* Edward Arnold, London

Smith D M 1987 Neoclassical location theory. In Lever W F (ed) *Industrial change in the United Kingdom.* Longman, Harlow, Essex: 23–37

Stabler H, Howe E 1988 Service exports and regional growth in the post-industrial era. *Journal of Regional Science* **28**: 303–15

Stanback T M 1980 *Understanding the service economy.* John Hopkins University Press, Baltimore

Stanback T M, Noyelle T J 1982 *Cities in Transition.* Rowman and Allanheld & Osman, Totowa, NJ

Stanback T M 1991 *The new suburbanization.* Westview Press, Boulder, CO

Stanback T M, Noyelle T J 1988 Productivity in services: a valid measure of economic performance. In Guile B, Quinn J B (eds) *Technology in services: policies for growth, trade and employment.* National Academic Press, Washington, DC: 187–211

Stanback T M, Bearse P J, Noyelle T J, Karesek R A 1981 *Services: the new economy.* Allanheld & Osmun, Totowa NJ

Starks M 1991 *Not for profit not for sale*, Reshaping the Public Sector, Vol. 6. Policy Journals, Newbury

Steaheli L A 1989 Accumulation, legitimation and the provision of public services in the American metropolis. *Urban Geography* **10**, 229–50

Stephens J D, Holly B P 1981 City systems behaviour and corporate influence: the headquarters location of US industrial firms, 1955–1975. *Urban Studies* **18**: 285–300

Stigler G J 1956 *Trends in employment in service industries.* National Bureau for Economic Research, Princetown University Press, New York

Stoker G 1989 Creating a local government for a post-Fordist society: the Thatcherite project. In Stewart J, Stoker G (eds) *The future of local government.* Macmillan, Basingstoke, Hants: 141–70

Stoker G 1991 Regulation theory, local government and the transition from Fordism. In King D, Pierre J (eds) *Challenges to local government.* Sage, London: 242–64

Stonier T 1983 *The wealth of information: a profile of the post-industrial economy.* Methuen, London

Storper M, Christopherson S 1987 Flexible specialisation and regional industrial agglomeration: the case of the US motion picture industry. *Annals of the Association of American Geographers* **77**: 104–17

Storper M, Walker R 1989 *The capitalist imperative.* Basil Blackwell, Oxford

Taylor W 1991 The logic of global business. *Harvard Business Review* **March/April:** 91–105

Thomas G, Miles I 1986 *Technology and changes in the provision of services to households,* FAST Occasional Papers No. **90**. Commission of the European Communities, Brussels

Thompson G F 1990 If you can't stand the heat, get off the beach: the UK holiday business. In Tomlinson A (ed) *Consumption, identity and style: marketing meanings and the packaging of pleasure.* Routledge, London: 195–220

Thorngren B 1970 How do contact systems affect regional development? *Environment and Planning A* **2**: 409–27

Thrift N J 1985 Taking the rest of the world seriously? The state of British urban and regional research in a time of economic crises. *Environment and Planning A.* **17**: 7–24

Thrift N J 1987 The fixers: the urban geography of international finance capital. In Henderson J, Castells, M (eds) *Global restructuring and territorial development.* Sage, Beverley Hills: 203–23

Thrift N J 1990a Doing regional geography in a global system: the new international financial system, the City of London and the South East of England. In Johnson R, Haver J, Hoekveld G A (eds) *regional geography: current developments and future prospects.* Routledge, Chapman & Hall, Andover, Hants: 180–216

Thrift N J 1990b The perils of the international financial system. *Environment and Planning A* **22**: 1135–7

Thrift N J 1994 On the social and cultural determinants of international financial centres: the case of the City of London. In Corbridge S, Mann R, Thrift N J (eds) Money, Power and Space. Basil Blackwell, Oxford: 327–55

Thrift N J, Leyshon A 1988 The gambling propensity: banks, developing country debt exposures and the view of the international financial system. *Geoforum* **19**: 55–69

Thrift N J, Leyshon A 1992 In the wake of money. In Budd L and Whimster S (eds) *Global finance and urban living. A study of metropolitan change.* Routledge, London: 282–311

Thurow L C 1989 Regional transformation and the service activities in Rodwin L, Sazanami H (eds) *Deindustrialisation and regional economic transformation: the experience of the United States.* Unwin Hyman, Boston, MA: 179–98

Tomlinson A 1990 Consumer culture and the aura of the commodity. In Tomlinson A (ed) *Consumption, identity and style: marketing meanings and the packaging of pleasure.* Routledge, London: 1–38

Tordoir P P 1991 Advanced office activities in the Randstad–Holland metropolitan region: location, complex formation and international orientation. In Daniels P W (ed) *Services and metropolitan development. International perspectives.* Routledge, Chapman & Hall, London: 226–44

Townsend A 1991 Services and local economic development. *Area* **23**: 309–17

Townsend A 1992 New directions in the growth of tourism employment? Propositions of the 1980s. *Environment and Planning A* **24**: 821–32

Tschetter J 1987 Producer service industries: why are they growing so rapidly? *Monthly Labor Review* **12**: 31–40

Tucker K, Sundberg M 1988 *International trade in services.* Routledge, London

United Nations Committee on Trade and Development (UNCTAD) **1989** *Trade and development report 1988.* United Nations, New York

Urry J 1986 Capitalist production, scientific management and the service class. In Scott A, Storper M (eds) *Production, work, territory.* Allen & Unwin, Winchester MA: 43–66

Urry J 1987 Some social and spatial aspects of services. *Environment and Planning D: Society and Space* **5**: 5–26

Urry J 1988 Cultural change and contemporary holidaymaking. *Theory, culture and society* **5**: 35–55

Urry J 1990a The 'consumption' of tourism. *Sociology* **24**: 23–35

Urry J 1990b *The tourist gaze: leisure and travel in contemporary societies.* Sage, London

Urry J 1990c Work, production and social relations. *Work, employment and society* **4**: 271–80

Urry J 1990d The service economy and the politics of place. Paper presented to the Pathways to Industrialisation and Regional Development in the 1990s. University of California, Los Angeles

Van Dinteren J H J 1987 The role of business services in the economy of medium-sized cities, *Environment and Planning A* **19**: 669–86

Vontoras N 1990 Emerging patterns of multinational enterprise operations in developed market economies: evidence and policy. *Review of Political Economy* **2**: 188–220

Walby S 1989 Flexibility and the changing sexual division of labour. In Wood S (ed) *The transformation of work* Unwin Hyman, London: 127–40

Walker R 1985 Is there a service economy? *Science and Society* **49**: 42–83

Walker R 1989 A requiem for corporate geography: new directions in industrial organisation, the production of places and uneven development. *Geografiska Annaler B* **71**: 43–68

Ward J 1991 *Tourism in action: 10 case studies in tourism*. Stanley Thornes, Cheltenham, Glos

Warde A 1985 Spatial change, politics and the division of labour. In Gregory D, Urry J (eds) *Social relations and spatial structures*. Macmillan, London: 190–214

Warde A 1990 Production, consumption and social change: reservations regarding Peter Saunder's sociology of consumption. *International Journal of Urban and Regional Research* **14**: 228–48

Warf B 1989 Telecommunications and the globalization of financial services. *Professional Geographer* **41**: 257–71

Warf B 1990 Deindustrialisation, service sector growth and the underclass in the New York metropolitan region. *Tijdschrift voor Economische en Sociale Geografie* **81**: 332–47

Warf B 1991 The internationalisation of New York services. In Daniels P W (ed) *Services and metropolitan development: international perspectives*. Routledge, Chapman & Hall, London: 245–64

Watts H D 1980 *Industrial geography*. Longman, Harlow, Essex

Weber A 1929 *Theory of the location of industries*. University of Chicago Press, Chicago

Werneke A 1983 *Microelectronics and office jobs: the impact of the chip on women's employment*. International Labour Office, Geneva

Westaway E J 1974 The spatial hierarchy of business organisations and its implications for the British urban system. *Regional Studies* **8**: 145–55

Wheeler J O 1985 The US metropolitan, corporate and population hierarchies, 1960–1980. *Geografiska Annaler B* **67**: 89–97

Wheeler J O 1986 Corporate spatial links with financial institutions: the role of the metropolitan hierarchy. *Annals of the Association of American Geographers* **76**: 262–74

Wheeler J O 1987 'Fortune' firms and the fortunes of their headquarters metropolises. *Geografiska Annaler B* **69**: 65–72

Wheeler J O 1988 The corporate role of large metropolitan areas in the United States. *Growth and Change* **19**: 75–86

Wheeler J O 1990 The new corporate landscape: America's fastest growing companies. *Professional Geographer* **42**: 433–44

Wheeler J O, Brown C L 1985 The metropolitan corporate hierarchy in the US South. *Economic Geography* **61**: 66–78

Wheeler J O, Dillon P M 1985 The wealth of the nation: the US metropolitan banking hierarchy. *Urban Geography* **6**: 297–315

Williams A M, Shaw G 1991 *Tourism and development: West European experiences*, 2nd Edn. Belhaven, London

Williams A M, Shaw G, Greenwood J 1989 From tourist to tourism entrepreneur, from consumption to production: evidence from Cornwall, England. *Environment and Planning A* **21**: 1639–53

Williams C C, Windebank J 1992 The implications for informal economic activity of European integration. Paper presented to the IBG Conference on 'An Integrating Europe', September, Birkbeck College, London

Williams C C, Windebank J 1993 Social and spatial inequalities in the informal economy: some evidence from the European Community. *Area* **25**: 358–64

Williams K, Williams J, Haslam C 1990 The 'hollowing out' of British manufacturing and its implications for policy. *Economy and Society* **19**: 456–88

Winckler V 1990 Restructuring the civil service: reorganisation and relocation. *International Journal of Urban and Regional Research* **14**: 135–57

Wood P A 1984 The regional significance of manufacturing–service sector links: some thoughts on the revival of London's docklands. In Barr B M and Waters N M (eds) *Regional diversification and structural change* Tantalus Research, Vancouver: 168–84

Wood P A 1986 The anatomy of job loss and job creation: some speculation on the role of the producer service sector. *Regional Studies* **20**: 37–40

Wood P A. 1987a Producer services and economic change: some Canadian evidence. In Chapman K, Humphrys G (eds) *Technological change and industrial policy*. Blackwell, London: 51–77

Wood P A 1987b Behavioural approaches to industrial location studies. In Lever W (ed) *Industrial change in the United Kingdom*. Longman, Harlow, Essex: 38–55

Wood P A 1988 The economic role of producer services; some Canadian evidence. In Marshall J N *et al., Services and Uneven Development*. Oxford University Press, Oxford: Section 9.3, 268–78

Wood P A 1991a Flexible accumulation and the rise of business services. *Transactions of the Institute of British Geographers* **16**: 160–77

Wood P A 1991b The Single European Market and producer service location in the United Kingdom. In Wild T, Jones P (eds) *De-industrialisation and new industrialisation in Britain and Germany*. Anglo-German Foundation, London

Wood P A, Bryson J, Keeble D 1993 Regional patterns of small firm development in business services: evidence from the UK. *Environment and Planning A* **25**: 677–700

Wood S (ed) 1989 *The transformation of work? Skill, flexibility and the labour process*. Unwin Hyman, London

Wrigley N 1987 The concentration of capital in UK grocery retailing. *Environment and Planning A* **19**: 1283–8

Wrigley N 1989 The lure of the USA: further reflections on the internationalisation of British grocery retailing. *Environment and Planning A* **21**: 283–8

Wrigley N 1991 Is the 'golden age' of British retailing at a watershed? *Environment and Planning A* **23**: 1537–44

Wrigley N 1993 Abuses of market power? Further reflection on UK food retailing and the regulatory state. *Environment and Planning A* **25**: 1545–52

References for the readings

Abler R, Falk T 1980 Intercommunications, distance, and geographical theory. *Geografiska Annaler Series B* **62**: 59–67

Abraham K 1988 *Flexible staffing arrangements and employers' short term adjustment strategies'* Working Paper **2619**. National Bureau of Economic Research, Cambridge, MA

Allen J 1988 The geographies of service. In Massey D, Allen J (eds) *Uneven redevelopment: cities and regions in transition.* Hodder & Stoughton, London: 124–41

Allwood P 1980 Putting the Companies Registration Office on microfilm. *Management Service in Government* **30**: 76–85

Altshuler A, Anderson A, Jones D, Roos D, Womack J 1984 *The future of the automobile.* MIT Press, Cambridge, MA

Appelbaum E 1985 Alternative work schedules of women. Paper prepared for the panel on technology and women's employment and related social issues. National Research Council, Washington, DC

Appelbaum E, Albin P 1988 *Employment, occupational structure and educational attainment in the United States, 1973, 1979 and 1987.* A report to the OECD, Commission on Services, Paris

Atkinson W R 1980 The employment consequences of computers: a user view. In Forester T (ed) *The microelectronics revolution.* Basil Blackwell, Oxford

Bacon R, Eltis W 1978 *Britain's economic problem: too few producers.* Macmillan, London

Bagnasco A 1977 *Tre Italie: la problematica dello Svillupo.* Il Mulino, Bologna

Bakis H 1985 *Telecommunication and organisation of company work space.* Centre National d'Etudes de Télécommunications, Paris

Baran B 1986 The technological transformation of white-collar work: a case study of the insurance industry. In Hartmann H, Kraut R, Tilly L (eds) *Technology and women's employment*, Vol. II. National Academy Press, Washington, DC

Barham J 1986 *Backstairs Cambridge.* Elison's Editions, Orwell, Herts

Baudrillard J 1981 *For a critique of the political economy of the sign.* Telos, St Louis

Berger S, Piore M 1980 *Dualism and discontinuity in industrial societies.* Cambridge University Press, New York

Berman L 1981 Towards a slimmer statistical service. *British Business* **27** February

Berry B 1970 The geography of the United States in the year 2000. *Transactions of the British Institute of Geographers* **51**: 21–54

Boddy M, Lovering J, Bassett K 1986 *Sunbelt city?* Clarendon Press, Oxford

Borchert J 1978 Major control points in American economic geography. *Annals of the American Association of Geographers* **62**: 214–32

Bourdieu P 1984 *Distinction: a social critique of the judgement of taste.* Routledge & Kegan Paul, Andover, Hants

Bourdieu P 1986 The production of belief: contribution to an economy of symbolic goods. In Collins R, Curran J, Garnham N, Scannell P, Schlesinger P, Sparks C (eds) *Media, Culture and Society.* Sage, Beverley Hills: 131–63

Bowlby S 1990 Technical change and the gender division of labour: the new information technology industries in Britain. *Geoforum* **21**(1): 67–84

Bowlby S R, Foord J, McDowell L 1986 The place of gender in locality studies. *Area* **18**: 327–31

Boyer R, Coriat B 1986 Technical flexibility and macro stabilization. Paper presented to the Conference on Innovation Diffusion, Venice, March

Braudel F 1967 *Capitalism and material life 1400–1800.* Harper & Row, New York

Brindle D 1987 *Financial Times* 26 November

British Telecom Unions Committee 1986 *A fault on the line: report on the first two years of privatised British Telecom.* British Telecom Unions Committee, Union Communications, 324 Grays Inn Road, London WC1X 8BX

Brusco S, Sabel C 1981 Artisanal production and economic growth. In Wilkinson F (ed) *The dynamics of labor market segmentation.* Academic Press, London: 99–113

Budd A 1978 *The politics of economic planning.* Fontana, London

Burns A 1934 *Production trends in the United States since 1870.* National Bureau of Economic Research, New York

Carre F 1988 *Temporary and contingent employment in the eighties, review of the evidence.* Economic Policy Institute, Washington, DC

CCC 1985a *Community welfare and development plan.* Cambridge County Council, Shire Hall, Cambridge CB3 OAP

CCC 1985b Tourism and employment in Cambridge, Cambridge County Council, Shire Hall, Cambridge CB3 OAP

CCC 1986 Employment development strategy: high tech and conventional manufacturing industry. Cambridge County Council, Shire Hall, Cambridge CB3 OAP

CCC 1987 Employment development strategy: the traded service sector in the Cambridge area. Cambridge County Council, Shire Hall, Cambridge CB3 OAP

Christopherson S, Storper M 1986 The city as studio, the world as back lot: the impact of vertical disintegration on the location of the motion picture industry. *Environment and Planning D: Society and Space* **4**: 305–20

Civil Service Commission 1966 *Annual Report.* HMSO, London

Civil Service Commission 1971 *Annual Report.* HMSO, London

Civil Service Commission 1979 *Annual Report.* HMSO, London

Civil Service Commission 1980 *Annual Report.* HMSO, London

Civil Service Commissioners 1962 *Annual Report, 1 January 1962 to 31 December 1962.* HMSO, London

Civil Service Department 1971a *Computers in central government ten years ahead*, Management Studies **2**. HMSO, London

Civil Service Department 1971b *The employment of women in the civil service*, Management Studies **3**. HMSO, London

Civil Service Department 1975 *Civil service statistics*. HMSO, London

Civil Service Department 1977 *Civil service statistics*. HMSO, London

Clark GL 1983 Fluctuations and rigidities in local markets. Part 1: theory and evidence. *Environment and Planning A* **15**: 165–85

Clark G 1986 The crisis of the midwest auto industry. In Scott A, Storper M (eds) *Production, work and territory*. Allen & Unwin, Hemel Hempstead, Herts

CMND 4506 1970 *The reorganisation of central government*. October

CMND 7057 1978 *Committee on the political activities of civil servants*. Chairman Sir Arthur Armitage, January

CMND 8504 1982 *Administrative forms in government*. Presented to parliament by the Chancellor of the Duchy of Lancaster, February

CMND 8616 1982 *Efficiency and effectiveness in the civil service: government observations on the third report from the Treasure and Civil Service Committee*. September

CMND 9058 1983 *Financial management in government departments*. September

Cockburn C 1985 *Machinery of dominance: women, men and technical know-how*. Pluto Press, London

Corby S 1984 Civil servant and trade union member: a conflict of loyalties? *Industrial Relations Journal* **15**(2): 18–29

Davies K, Gilligan C, Sutton C 1985 Structural changes in grocery retailing: the implications for competition. *International Journal of Physical Distribution and Material Management* **15**: 3–48

De Neubourg G 1985 Part-time work: an international quantitative comparison. *International Labour Review* **124**: 559–76

Ehrenberg R, Rosenberg P, Li J 1986 Part-time employment in the United States. Paper presented at a conference on Employment, Unemployment and Hours of Work, Berlin, Germany, September

Estall R C 1985 Stock control in manufacturing. The just-in-time system and its locational implications. *Area* **17**(2): 129–33

Evans C 1985 Privatisation of local services. *Local Government Studies* **11**: 97–110

Ewen S 1976 *Captains of consciousness: advertising and the social roots of consumer culture*. McGraw-Hill, New York

Ewen S 1988 *All consuming images: the politics of style in contemporary culture*. Basic Books, New York

Featherstone M 1983 Consumer culture: an introduction. *Theory, Culture and Society*. **1**(3): 4–9

Ferguson D, Haycraft D, Segal N 1987 *Cambridge*. Covent Garden Press, Cambridge

Frampton K 1983 Towards a critical regionalism: six points for an architecture of resistance. In Foster H (ed) *Postmodern culture*. Pluto Press, London: 16–56

Fulton Report 1968 The civil service, Vols 1–4. CMND 3638

Galtung J 1982 The new international order: economics and communication. In Jussawalla M, Lamberton D (eds) *Communication economics and development*. Pergamon, New York: 133–43

Gamble A 1981 *Britain in decline*. Macmillan, London

Gershuny J I, Miles I D 1983 *The new service economy: the transformation of employment in industrial societies*. Frances Pinter, London

Giddens A 1984 *The constitution of society*. Polity Press, Cambridge

Goddard J 1975 *Office location in urban and regional development*. Oxford University Press, Oxford

Goddard J 1980 Technological forecasting in a spatial context. *Futures* **12**: 90–105

Goddard J, Pye R 1977 Telecommunications and office location. *Regional Studies* **11**: 19–30

Goddard J, Gillespie A, Robinson F, Thwaites A 1985 The impact of new information technology on urban and regional structure in Europe. In Thwaites A, Oakey R (eds) *The regional impact of technological change* Frances Pinter, London: 215–41

Goffman E 1959 *The presentation of self in everyday life*. Doubleday, New York

Gough I 1979 *The political economy of the welfare state*. Macmillan, London

Gyford J 1983 The new urban left: a local road to socialism. *New Society* **21 October**: 91–3

Haber S, Lamas E, Lichtenstein J 1987 On their own: the self-employed and others in private business. *Monthly Labor Review* **110**: 17–23

Hall P 1985 The geography of the fifth Kondratieff. In Hall P, Markusen A (eds) *Silicon landscapes*. George Allen & Unwin, New York: 1–19

Hammond E 1967 Dispersal of government offices: a survey. *Urban Studies* **4**: 258–75

Hammond E 1968 *London to Durham – transfer of the Post Office Savings Certificate Division*. University of Durham Rowntree Research Unit, Durham

Hardman Report 1973 *The dispersal of government work from London* CMND 5322, June

Helleiner G, Cruise O'Brien R 1982 The political economy of information in a changing international economic order. In Jussawalla M, Lamberton D (eds) *Communication economies and development*. Pergamon, New York: 100–32

Hepworth M 1985 Geography of the information economy. Doctoral dissertation, Department of Geography, University of Toronto

Hewison R 1987 *The heritage industry*. Methuen, London

HM Treasury 1984a *Civil service statistics 1984*. HMSO, London

HM Treasury 1984b A smaller civil service. *Economic Progress Report* **168**: 1–8

HM Treasury 1984c Computerisation of Pay-As-You-Earn. *Economic Progress Report* **170**: 6–7

Hochschild A R 1983 *The managed heart: commercialisation of human feeling*. University of California Press, Berkeley, CA

Honey R, Sorenson D 1984 Jurisdictional benefits and local costs: the politics of school closings. In Kirby A, Knox P, Pinch S (eds) *Public service provision and urban development*. Croom Helm, Beckenham, Kent: 114–30

House of Commons Committee of Public Accounts 1981 *The control of civil service manpower.* HC 367–i, 1980–81

House of Commons Committee of Public Accounts 1984 *Manpower control – reviewing the need for work.* HCISO, 1983–84

House of Commons Committee on Scottish Affairs 1980 *Dispersal of civil service jobs to Scotland.* HC 88, 1980–81

House of Commons Debates 18 July 1963, Volume 681, Written Answers Columns 82–6

House of Commons Debates 4 June 1965, Volume 715, Columns 2176–92

House of Commons Debates 19 December 1971, Volume 827, Columns 1497–8

House of Commons Debates 30 July 1974, Volume 878, Columns 482–94

House of Commons Debates 26 July 1979, Volume 971, Columns 902–22

House of Commons Treasury and Civil Service Committee 1981 *Efficiency and effectiveness in the civil service.* HC 360–i 1980–81

House of Commons Treasury and Civil Service Committee Sub-Committee 1981 *Efficiency and effectiveness in the civil service,* HC 111, 1981–82

Huckfield L 1978 Business Statistics Office: what it does and why. *Trade and Industry* **31**: 483–94

Ichnowski B, Preston A 1985 *New trends in part-time employment.* Proceedings of the 38th Annual Meeting of the Industrial Relations Research Association

Ikeda M 1979 The subcontracting system in the Japanese electronic industry. *Engineering Industries of Japan* **19**: 43–71

Jager M 1986 Class definition and the aesthetics of gentrification: Victoriana in Melbourne. In Smith N, Williams P (eds) *Gentrification of the city.* Allen & Unwin, Boston, MA. 78–91

Jameson F 1983 Postmodernism and consumer society. In Foster H (ed) *The anti-aesthetic.* Bay Press, Port Townsend, WA: 111–25

Jameson F 1984 Postmodernism, or the cultural logic of late capitalism. *New Left Review.* **146** 52–92

Jencks C 1987 *The language of post-modern architecture*, 5th ed. Academy Editions, London

Jhally S 1987 *The codes of advertising*. Frances Pinter, London

Joint Review Group on Employment for Women in the Civil Service 1982 *Equal opportunities for women in the civil service*. Treasury Management and Personnel Office. HMSO, London

Karan T 1984 The local government workforce – public sector paragon or private sector parasite. *Local Government Studies* **10**: 39–58

Keeble D E 1989 High technology industry and regional development in Britain: the case of the Cambridge phenomenon. *Environment and Planning C: Government and Policy* **7**: 153–72

Kowinski W S 1985 *The malling of America: an inside look at the great consumer paradise*. William Morrow, New York

Kuznets S 1930 *Secular movements in production and prices*. Houghton-Miffle, Boston, MA

Labour Research Department 1984 *Bargaining report on women*. LRD Publications, 78 Blackfriars Road, London SE1 8HF

Langdale J 1979 The role of telecommunications in information economy. *Seminar on Social Research and Telecommunications Policy, Background Papers*, Telecom Australia

Lasch C 1979 *The culture of narcissism: American life in an age of diminishing expectations*. Warner Books, New York

Le Grand J 1984 The future of the welfare state. *New Society* **7 June**: 385–6

Leon C, Bednarzik R 1978 A profile of women on part-time schedules. *Monthly Labor Review* **101**: 3–12

Loveman G, Sengenberger W 1988 *The reemergence of smaller units of employment, developments of the small and medium-sized enterprise sector in industrialized countries*. International Labour Organization, Geneva

McCracken G 1988 *Culture and consumption: new approaches to the symbolic character of consumer goods and activities*. Indiana University Press, Bloomington

McLuhan M 1964 *Understanding media*. Routledge & Keegan Paul, London

Marshall G 1986 The workplace culture of a licensed restaurant. *Theory, Culture and Society* **3**(1): 33–47

Martin R L 1986 In what sense a jobs boom? Employment recovery, government policy and the regions. *Regional Studies* **20**: 463–72

Massey D 1984 *Spatial divisions of labour*. Macmillan, London

Massey D 1988 Uneven development: social change and spatial divisions of labour. In Massey D, Allen J (eds) *Uneven redevelopment: cities and regions in transition*. Hodder & Stoughton, London: 250–76

Miller M B 1981 *The bon marche: bourgeois culture and the department store, 1869–1920*. Princeton University Press, Princeton

Mohan J 1988a Spatial aspects of health-care employment in Britain: 2. Current policy initiatives. *Environment and Planning A* **20**: 203–17

Mohan J 1988b Spatial aspects of health-care employment in Britain: 1. Aggregate trends. *Environment and Planning A* **20**: 7–23

Moir C 1990 Competition in the UK grocery trades. In Moir C, Dawson J (eds) *Competition and markets: essays in honour of Margaret Hall*. Macmillan, London: 91–118

Monopolies and Mergers Commission 1981 *Discounts to retailers*. HC 311. HMSO, London

Moon G, Parnell R 1986 Private sector involvement in local authority service delivery. *Regional Studies* **20**: 253–66

Moore B, Spires R 1986 The role of high technology complexes and science parks in regional development. Paper presented at the OECD Seminar on Science Parks and Technology Complexes in Relation to Regional Development, Venice, 3 June. Copy available from B Moore, Department of Land Economy, University of Cambridge, Cambridge

Morse D 1969 *The peripheral worker*. Columbia University Press, New York

Murray R 1988 *Crowding out: boom and crisis in the South East*. South East Economic Development Strategy, Stevenage, Herts

National Research Bureau 1990 *Shopping center directory 1990*. NRB, Chicago

Nilles J, Carlson F, Gray P, Hanneman G 1976 *The telecommunications–transportation trade-off*. Wiley Interscience, New York

Noyelle T 1986 *Beyond industrial dualism: market and job segmentation in the new economy.* Westview Press, Boulder, CO

O'Connor J 1973 *The fiscal crisis of the state.* St Martins Press, New York

Office of Fair Trading 1985 *Competition and retailing.* Office of Fair Trading, Field House, Breams Buildings, London EC4A 1PR

Organization for Economic Cooperation and Development 1986 *Flexibility in the labour market: a technical report.* OECD, Paris

Owen K 1974 *Computing in government.* Civil Service Department, London

Patterson A 1987 The restructuring of water production and consumption. Paper presented at the Institute of British Geographers Annual Meeting, Portsmouth Polytechnic, January. Available from Department of Land Management, University of Reading, Reading, Berks

Paul J 1984 Contracting out in the NHS: can we afford to take the risk? *Critical Social Policy* **10**: 87–94

Peters T J, Waterman R H 1982 *In search of excellence: lessons from America's best businesses.* Harper & Row, New York

Pinch SP 1987 The changing geography of preschool services in England between 1977 and 1983. *Environment and Planning C: Government and Policy* **5**: 469–80

Piore M, Sabel C 1984 *The second industrial divide* Basic Books, New York

Plewes, T 1987 Understanding the data on part-time and temporary employment. Paper presented at a conference on Women and the contingent workforce, sponsored by the US Department of Labor, Women's Bureau, New York, February

Pool I de S (ed) 1977 *The social impact of the telephone.* MIT Press, Cambridge, MA

Pred A 1977 *City systems in advanced economies.* Wiley, New York

Privy Council 1981 *Efficiency in the civil service.* CMND 8293, July

Relph E 1987 *The modern urban landscape.* Johns Hopkins University Press, Baltimore

Richardson H, Clapp J 1984 Technological change in information processing industries and regional income differentials in developing countries. *International Regional Science Review* **9**: 241–56

Rose H, Rose S 1982 Moving right out of welfare – and the way back. *Critical Social Policy* **2** (summer issue): 7–18

Rothwell R 1984 Technological innovation and long waves in economic development. Seminar on Cambio Technologico j Desarollo Economico, University Internaçional Menendez Pelayo, Santander, Spain

Sack D 1988 The consumer's world: place as context. *Annals of the Association of American Geographers* **78**(4): 624–64

Sahlins M 1976 *Culture and practical reason*. Chicago University Press, Chicago

Sayer A 1985 Industry and space: a sympathetic critique of radical research. *Society and Space* **3**: 3–30

Sayer A, Morgan K 1986 The electronics industry and regional development in Britain. In Amin A, Goddard J B (eds) *Technological change, industrial restructuring and regional development*. Allen & Unwin, London: 157–87

Schoen C 1987 Testimony before a hearing of the US National Labor Relations Board in relation to Healthcare Bargaining Units, October 14

Schumpeter J 1942 *Capitalism, socialism, and democracy*. Harper & Row, New York

Scott A J 1981 The spatial structure of metropolitan labor markets and the theory of intra-urban plant location. *Urban Geography* **2**: 1–30

SEEDS 1987 *South-South divide*. South East Economic Development Strategy, Stevenage, Herts

Semple R, Green M 1983 Interurban corporate headquarters relocation in Canada. *Cahiers de Geographie du Quebec* **27**: 389–406

Sharrett C 1989 Defining the postmodern: the case of SoHo Kitchen and El Internacional. In Kellner D (ed) *Postmodernism/Jameson/Critique*. Maisonneuve Press, Washington DC: 162–71

Sheftner M 1980 New York fiscal crisis: the politics of inflation and retrenchment. In Levine C (ed) *Managing fiscal stress*. Chatham House, Chatham, NJ: 251–72

Shopping Center Age 1989 But is it art? Museum, living or 'plop' – artwork is good business for malls. *Shopping Center Age* **November**: 104–12

Small Business Administration 1987 *The state of small business: a report of the president.* US GPO, Washington, DC

Spence M, 1981 Signalling, screening, and information. In Rosen S (ed) *Studies in labor markets.* University of Chicago Press, Chicago: 319–57

Spencer K 1984 Assessing alternative forms of service provision. *Local Government Studies.* **10**: 14–20

SQP 1985 *The Cambridge phenomenon: the growth of high technology in a university town.* Segal Quince and Partners, Mount Pleasant House, Mount Pleasant, Cambridge, UK

Stanback T 1987 *Computerization and the transformation of employment.* Westview Press, Boulder, CO

Standing G 1986 *Unemployment and labour market flexibility: the United Kingdom.* International Labour Organization, Geneva

Stigler G J 1961 The economics of information. *Journal of Political Economy* **69**: 213–55

Stigler G J 1962 Information in the labor market. *Journal of Political Economy* **70**: 94–105

Stoffel J 1988 What's new in shopping malls. *New York Times,* 7 August

Stone K 1981 The post war paradigm in American labor law. *Yale Law Journal* **90**: 1509–80

Storper M 1985 Oligopoly and the product cycle: essentialism in economic geography. *Economic Geography* **61**(3): 260–82

Storper M, Christopherson S 1985 *The changing location and organization of the motion picture industry: interregional shifts in the United States.* UCLA Graduate School of Architecture and Urban Planning, Research Report no. 127, Los Angeles.

Tanenbaum A 1981 *Computer networks.* Prentice Hall, Englewood Cliffs, NJ

Thorgren B 1970 How do contact systems affect regional development? *Environmental Planning* **2**: 409–27

Tilly C 1988 *Short hours, short shrift, understanding part-time employment.* Economic Policy Institute, Washington, DC

Tornqvist G 1977 Comment. In Ohlin B, Hesselborn P, Wyjkman P, *The international allocation of economic activity.* Macmillan, New York: 445–51

Travers T 1983 Local government de-manning. *Public Money* **3** (June): 64–7

Urban Land Institute 1985 *Shopping center development handbook*. Urban Land Institute, Washington DC

Urry, J 1986 *Services, some issues of analysis*. WP–17, Lancaster Regionalism Group, University of Lancaster, Lancaster

Urry J 1987 Some social and spatial aspects of services. *Environment and Planning D: Society and Space* **5**: 5–26

US Department of Commerce 1972 *Census of retail trade*. US GPO, Washington, DC

US Department of Commerce 1977 *Census of retail trade*. US GPO, Washington, DC

US Department of Commerce 1982 *Census of retail trade*. US GPO, Washington, DC

US Department of Commerce, Bureau of the Census 1983 *Current population survey*

Vernon R 1966 International investment and international trade in the product cycle. *Quarterly Journal of Economics* **80**(2): 190–207

Warde A 1986 Industrial restructuring, local politics and the reproduction of labour power: some theoretical considerations. *Environment and Planning D: Society and Space* **6**: 75–96

Waterman D 1982 The structural development of the motion picture industry. *American Economist* **26**(1): 16–27

Webster B 1985 A women's issue: the impact of local authority cuts. *Local Government Studies* **11**: 19–46

Wessex Regional Health Authority 1987 *Annual Report 1986*. Wessex Regional Health Authority, Highcroft, Romsey Road, Winchester, Hants

Whyte E F 1948 *Human relations in the restaurant industry*. McGraw-Hill, New York

Williams R 1980 *Problems in materialism and culture: selected essays*. New Left Books, London

Williams R H 1982 *Dream worlds: mass consumption in the late nineteenth century France*. University of California Press, Berkeley, CA

Winckler V L 1986 Class and gender in regional change: a study of office relocation in industrial south Wales. Unpublished PhD thesis, University of Wales Institute of Science and Technology, Cardiff

Wrigley N 1989 The lure of the USA: further reflections on the internationalisation of British grocery retailing capital. *Environment and Planning A* **21**: 283–8

Index